Graduate Texts in Contemporary Physics

Series Editors:

R. Stephen Berry
Joseph L. Birman
Jeffrey W. Lynn
Mark P. Silverman
H. Eugene Stanley
Mikhail Voloshin

Springer
New York
Berlin
Heidelberg
Barcelona
Hong Kong
London
Milan
Paris
Singapore
Tokyo

Graduate Texts in Contemporary Physics

T. S. Chow

Mesoscopic Physics of
Complex Materials

With 83 Illustrations

 Springer

T. S. Chow
Xerox Research and Technology
800 Phillips Road, 0114-39D
Webster, NY 14580
USA
tchow@ctr.xerox.com

Series Editors

R. Stephen Berry
Department of Chemistry
University of Chicago
Chicago, IL 60637
USA

Joseph L. Birman
Department of Physics
City College of CUNY
New York, NY 10031
USA

Jeffrey W. Lynn
Department of Physics
University of Maryland
College Park, MD 20742
USA

Mark P. Silverman
Department of Physics
Trinity College
Hartford CT 06106
USA

H. Eugene Stanley
Center for Polymer Studies
Physics Department
Boston University
Boston, MA 02215
USA

Mikhail Voloshin
Theoretical Physics Institute
Tate Laboratory of Physics
The University of Minnesota
Minneapolis, MN 55455
USA

Library of Congress Cataloging-in-Publication Data
Chow, Tsu-sen.
    Mesoscopic physics of complex materials/Tsu-sen Chow.
    p.   cm. — (Graduate texts in contemporary physics)
    Includes bibliographical references and index.
    ISBN 0-387-95032-X (alk. paper)
    1. Mesoscopic phenomena (Physics).   I. Title.   II. Series.
QC176.8.M46 C47 2000
530.4'17–dc21   00-030464

Printed on acid-free paper.

Production managed by Frank McGuckin; manufacturing supervised by Jacqui Ashri.
Typeset by Techbooks, Fairfax, VA.
Printed and bound by Edwards Brothers, Inc., Ann Arbor, MI.
Printed in the United States of America.

9 8 7 6 5 4 3 2 1

ISBN 0-387-95032-X   SPIN 10764965

Springer-Verlag   New York   Berlin   Heidelberg
A member of BertelsmannSpringer Science+Business Media GmbH

Preface

This book is intended to provide a cross-disciplinary study of the physical properties of complex fluids, solids, and interfaces as a function of their mesoscopic structures. Because of the disorder and dissipate nature of these structures, emphasis is placed on nonequilibrium phenomena. These phenomena are the active research areas of soft condensed matter, and it is impossible to cover them all in one book. Therefore, we have limited the scope by selecting a variety of important current systems that (1) present high values to both science and technology on the basis of my own preference and expertise and (2) have not been put together coherently in the form of a book. We then show the underlying connections and parallels between topics as diverse as critical phenomena in colloidal dynamics, glass state relaxation and deformation, reinforced polymer composites, molecular level mixing in nanocomposites, and microscopic interactions of rough surfaces and interfaces. At the same time, each chapter is designed to be directly accessible to readers, and the need for going through the previous chapters has been kept to the minimum.

It is a reasonably short book that is not designed to review all of the recent work that spans many disciplines. Instead, we attempt to establish a general framework for the fundamental understanding and the practical development of new materials that cannot be designed by the trial-and-error methods. Statistical dynamics put greater emphasis on the time factor and is most suitable for the purpose of describing the dissipative and irreversible behavior of complex materials. Both the time and length scales are of principal interests.

The length scales are between atoms and microns. We shall put more emphasis on the basic models and concepts that are shared among different chapters for the simplicity and better physical understanding. Mathematical symbols proved a problem during the preparation of the manuscript, because a variety of subjects are being addressed. We have aimed at uniformity but sometimes were compelled to specify them for different usage in a different topic.

Although the purpose of this monograph is to introduce readers with theoretical methods and concepts in the applications of nonequilibrium statistical mechanics to complex materials, over 80 diagrams illustrate key points and compare theories and experiments. A familiarity with the basics of statistical mechanics and condensed matter physics is assumed. Of course, many detailed books on the specific topics of physics and soft materials are available, and some of them are going to be mentioned in the succeeding chapters. However, we are not aware of any book whose author has attempted the scope presented here. The book is written pedagogically. It is our intent that the depth and breadth of coverage will be useful as a graduate-level text in academia or a reference in industry. We would like to connect graduate students and researchers in condensed matter physics, polymer physics, colloid and interface science, materials science, and engineering to some important areas of modern materials science.

The author would like to take this opportunity to thank many colleagues over the years, starting from Professor J. J. Hermans to Dr. C. B. Duke for fruitful collaborations or valuable discussions. Individual acknowledgements for the works behind the illustrations and presentations are found in the list of references. To his wife Shang-mei, the author is indebted for her long and persevering encouragement and support.

Webster, NJ   T. S. Chow
June 2000

Contents

1
Overview

The best way to understand the guiding physical principles common to different aspects of complex materials is by covering as many examples as possible. Complex materials either refer to colloidal dispersions in the liquid state or to polymers and composites in the solid state, which give an impulse to the importance of rough surfaces and interfaces. In these materials, particles or molecules are organized into structures with the length scales between atoms and microns. They play a prominent role in so many high-technology applications. From the scientific point of view, they have been widely studied in the connection with disordered systems, mesoscopic physics, and soft condensed matter. Disorder is characteristic of complex materials, and it can be structural, compositional, or topological disorder. The physics of complex materials is a tremendously rich subject, and it is too late to capture all of the richness of their properties within a single volume. This book places its main emphasis on the nonequilibrium behavior in its relationships to disordered structures that are of great theoretical and experimental interest. A coherent physical picture is expected to emerge that shows the underlying connections and parallels between different disordered systems varying from liquid to solid, and from surface to interface. Although we shall deal mostly with theories, the essential experimental verifications of the theoretical calculations will be discussed.

Among various dynamic properties of complex materials, an important and conspicuous property is the viscoelasticity (see Appendix 1A), which is useful in characterizing the properties of fluid flow as well as solid deformation; it consists of contributions from the elastic storage part and the viscous dissipate part. This distinctive property will be focused in this monograph in the study of the dissipative and irreversible behavior of various disordered systems at deferent phases. By the way, viscoelasticity has also been a subject of intense investigation in the excellent books on the dynamics of polymeric liquids [1,2]. One of the main purposes of this book is to understand the macroscopic properties as a function of microscopic fundamentals through the methods and concepts developed in nonequilibrium statistical mechanics. Both the time and length scales are going to play important roles in the theoretical development. Mesoscopic physics help us to understand the macroscopic limits by building larger and larger length scales that go from molecular to macroscopic levels.

1.1   Statistical Dynamics

When a force is applied to a system or is removed from it after having kept it on for a time, the time-dependent response can best be addressed within the framework of statistical dynamics. It offers a powerful and versatile theoretical tool for the investigation of nonequilibrium processes in which the temporal evolution is explicitly considered. It is regarded as stochastic processes to be discussed in chapters 2 and 3. We start by introducing Brownian motion, which is basic to stochastic processes and nonequilibrium statistical mechanics [3,4], and that introduction sets the foundation for analyzing the dissipative and irreversible behavior of disordered systems. The linear response theory connects the response caused by the external disturbance to the spontaneous fluctuations that is related to the intrinsic fluctuations and cannot be switched off. The fluctuation–dissipation theorem is at the heart of relating the time-dependent properties to the spontaneous fluctuations expressed by correlation functions of a thermally equilibrium system.

All disordered systems possess a number of general properties in common. The most important ones are the spatial homogeneity in the mean and the absence of any correlation between the parameters that characterize the disorder at points that are far apart. When particles end up far from each other as the result of diffusion, the correlation carried by particles can be forgotten as time goes on. We thus lost information, and the entropy of the system increases with time. The linear response theory alone is not going to be sufficient to solve the problem. To reach the macroscopic level observation from the microscopic level, we need successive levels of coarse graining in the determination of the effective properties of complex materials. In the process, a certain amount of information gets lost in the probabilistic description of the problem, which is the fundamental theme of statistical physics [5,6,7]. We treat a few stochastic equations that include the Langevin equation, hydrodynamic fluctuations [8], and the Fokker–Planck equation, and finally lead to the master equation. These equations are not completely independent of each other and will be used throughout the book.

Fractals have become an increasingly useful tool in the study of disordered systems [9,10]. We shall use the fractal concepts in the development of scaling theories and in the discussion of the effects of disorder on the mesoscopic scales. In sections 4.10 and 4.11, scaling laws for chain networks are derived on the basis of a colloid growth model. A fractal dynamic theory of glasses is introduced in Section 5.3. The noise and fluctuations of self-affine fractal surfaces serve as the basis of discussions in Chapter 8.

1.2   Fluid Dispersions

Colloidal dispersions have interesting and complex flow behavior because the macroscopic properties are sensitive to the nonequilibrium microstructure, which in turn depends on the composition, flow field, and particle interaction to be discussed in Chapter 4. Both the time and length scales play important roles in the many-particle problems. Experimental data of the effective shear viscosity have exhibited several interesting and unusual features, and they serve as the good basis of theoretical discussions, especially in the case of concentrated dispersions, in which the cooperative phenomena are observed.

Probably the single, most important characteristic of concentrated dispersions is that they have a shear-rate–dependent [or frequency (ω)] viscosity. The shear viscosity (η) is Newtonian for dilute suspensions, and it becomes non-Newtonian for semidilute suspensions. The former is shear-rate independent, and the latter is shear-rate dependent. The phenomenon of decreasing the effective shear viscosity with an increase in shear rate occurring at the higher volume fraction (ϕ) of colloidal particles is known as shear thinning. It becomes more pronounced as ϕ approaches a percolation threshold ϕ_c generally believed to result from shear-induced change in microstructure. This unusual critical phenomenon has been observed for colloidal dispersions, but not for polymer solutions, even though polymer liquids also exhibit non-Newtonian flow behavior [1,2]. A better understanding of the many-body interactions between the short-range colloidal forces and the equilibrium microstructure will be presented. Let us see a few interesting examples of what we have just mentioned.

The zero-shear viscosities, $\eta(\phi, \omega \to 0)$, of both neutral and charged hard spheres are shown in Figure 1.1. Because of the electrostatic repulsion, the percolation

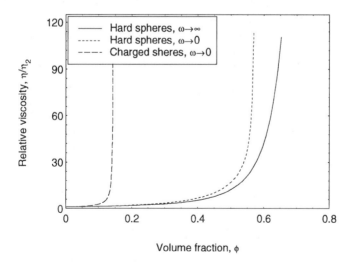

FIGURE 1.1.   The critical phenomena of the effective shear viscosity of hard-sphere colloids as a function of the frequency (ω) and the percolation threshold (ϕ_c) that depends on the repulsive potential.

transition for a charged particle at low ionic strength has a much lower value of ϕ_c. This shift in the critical phenomenon is related to the particle interaction. In Figure 1.1, we also see the difference between the low and high shear-rate limits that is related to the microstructural relaxation. In Chapter 4, the effects of particle size, shape, and orientation are also going to be analyzed. Another interesting example is to explore the viscoelastic nature of polymer gels far from the sol–gel transition as a function of the microstructure and particle interactions, which remain a topic of current research in complex fluids [11].

1.3   Relaxation in Solids

The disorder of amorphous solid is the result of the nonequilibrium nature of such a structure, and its time of existence is usually very long. Structural relaxation involves the time-dependent behavior of a system changing from one quasi-equilibrium state to another. It occurs in liquids as well as in solids; however, the focus in chapters 5 to 7 will be on the glassy-state relaxation of amorphous polymers and composites. A stronger temperature and time dependence of the physical properties of solid polymers exist, compared with those of other materials, such as metal and ceramics, because the glass transition temperature (T_g) of polymers is lower and the relaxation time is far more sensitive to temperature and time scale (deformation rate, cooling rate, etc.). A fractal dynamic theory of glasses will be presented to describe the structural relaxation and deformation kinetics of polymeric glasses and particulate composites. The concept of free volume (hole) has played a central role in discussing the physical properties of amorphous materials. The glassy-state relaxation is a result of the local configuration rearrangements of molecular segments, and dynamics of holes provide a quantitative description of the segmental mobility. On the basis of the dynamics of hole motion, a unified physical picture has emerged that enables us to analyze the equation of state (PVT), the glass transition, and viscoelasticity.

The transition of the relaxation time (τ) near the glass transition is shown in Figure 1.2 as a function of temperature (T) and cooling rate (q). It shows the departure from the well-known William–Landel–Ferry (WLF) equation [12] for $T < T_g$. Here, $a(T, q) \equiv \tau(T, q)/\tau_r$ is the relaxation time scale often called the shift factor and τ_r is a constant. This figure shows that T_g is a nonequilibrium property and the activation energy (ΔH) in the vicinity of T_g changes in accordance with

$$\Delta H(T < T_g) = [1 - \mu(T)] \cdot \Delta H(T > T_g), \tag{1.1}$$

where μ is the physical aging exponent [13] and depends on temperature. The effect of thermal history on glasses is included in its properties: $\mu(T) < 1$ for $T < T_g$, and $\mu = 0$ for $T > T_g$. This exponent plays an important role in the glassy-state deformation. Experimentally, the shift factor is obtained by the superposition of many curves of similar shape measure at different temperatures

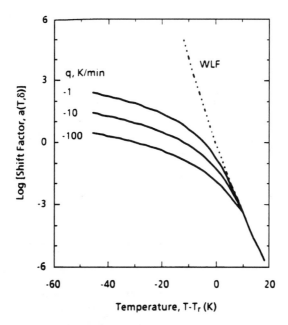

FIGURE 1.2. The departure of the relaxation time scale (a) from the WLF dependence in the vicinity of the glass transition temperature for a polymer such as polystyrene that is vitrified from liquid to glass at different cooling rates q.

and thermal history. These curves may be the plots of creep compliance versus time. The existence of such superposition suggests an inherent simplicity hidden behind the apparent complexity. It is this feature that we shall elaborate on.

When the stress level approaches plastic yielding, the motion of polymer segment is no longer local because the domain of molecular segments that has to move as a whole becomes much larger than the volume of a single lattice site. Different external stresses produce different volumes of the chain segments, and these volumes are the components of an activation volume tensor. This key parameter is needed for describing the nonlinear relaxation time. Figure 1.3 shows that the nonlinear stress-strain relationships of polymers and composites depend on the time scales in which the solids are measured and relaxed. In addition to temperature and the time-dependent glassy state (δ), the nonlinear structural relaxation is also a function of the stress tensor (σ_{ij}), and the volume fraction of filler (ϕ) in composites.

1.4 Nanocomposites

Nanostructures are going to be discussed throughout the book in the context of mesoscopic physics that links the macroscopic flow and deformation to the

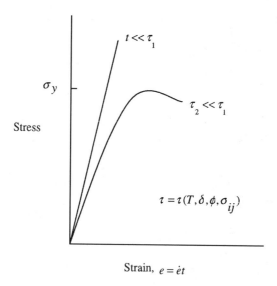

FIGURE 1.3. Dependence of the nonlinear stress–strain relationships on the relaxation times τ. The change of the deformation mechanism from ductile to brittle depends the time scales in which a glassy polymer is measured (t) and relaxed (τ). The strain rate is $\dot{e}$.

microscopic structures and interactions. The length scales are larger than those in the atomic range (a fraction of a nanometer) but smaller than microns. This subject has attracted current interests. Because this is an emerging area of research, we shall make only a few remarks about the unusual phenomena of nanocomposites with the particle sizes close to 10 nanometers or less that is in the range of molecular-level mixing.

Most composite theories are developed for reinforced polymeric composites in which the purpose has been to try to find ways of achieving very high elastic constants and tensile strength, and they are based on the micromechanics or strain energy calculations. Sections 7.6 to 7.8 demonstrate the limitation of the traditional composite theories, and they point out the importance of microscopic interactions and molecular parameters. We shall see the anomaly in the yield stress of compatible polymer blends and the order–disorder transition in polymer-diluent systems.

Sections 4.1 to 4.4 give another example of the observed effective viscosity of concentrated dispersions that cannot be explained by theories based on the continuum mechanics, which underestimate the concentration dependence. Therefore, the short-range molecular interactions within the nanostructure in the liquid state have to be analyzed in accordance with lattice models. They are basically theories of solids, and they have the advantage of providing a good connection between the dense liquids and amorphous solids [14]. Sections 4.8 to 4.9 reveal why the Newtonian flow behavior has been recently observed for fluids dispersed with particles having their sizes at 10 nanometers or less at high concentrations.

1.5   Fractal Surfaces

Rough surfaces and interfaces are ubiquitous in nature and, at the same time, play in ever more critical technology applications, such as microelectronics, image formation, coating and growth of thin films, etc. Therefore, it becomes increasingly important to understand the structure-property relationships of rough surfaces and interfaces. Their study requires us to consider a description of their mesostructure and an understanding of their properties: statics and dynamics. The statistical techniques alone are not adequate for describing a important class of rough surfaces observed experimentally, and the fractal geometry [15] is also needed as an essential part of the discussions in Chapter 8.

Many rough surfaces have the characteristic of a self-affine fractal that is statistically invariant under anisotropic dilation. The irregular fluctuations reflect the microstructure of a rough and can be analyzed as stochastic processes expressed by Brownian motion. The microstructure refers to the standard deviation (σ), correlation length (ξ), and the roughness exponent (α) that defines the scaling properties of the surface. The height correlation function $C(r)$ shown in Figure 1.4 reveals that the surface microstructure is short ranged. Macroscopic limit is usually approached once the size of the system is much greater than the correlation length ξ, which will be pivotal in describing the structure-property relationships of surfaces and interfaces.

Both the long-range noise correlation function and the height correlation function are going to be needed in the study of the effects of roughness on the wetting

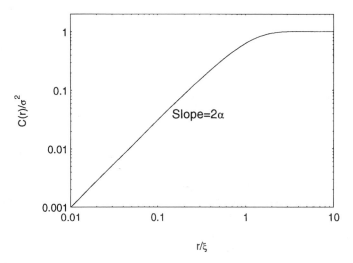

FIGURE 1.4. A double logarithmic plot of the height correlation function: The scaling behavior $C(r) \sim r^{2\alpha}$ in the range of $r < \xi$, the heights become uncorrelated and $C(r) = \sigma^2$ for $r > \xi$. The crossover fixes both correlations lengths σ and ξ that are normal and parallel to the surface, respectively.

and adhesion, the contact line depinning, the critical surface tension, and the transition from the partial wetting to the complete wetting. Beyond statics, we shall investigate the dynamics of wetting, the microscopic friction of surfaces in relative motion, and the diffusive light scattering. The surface adhesion and the bulk deformation are the two main contributions to friction that will be discussed. Finally, the concept of dynamic scaling will be used to explore the stochastic growth and fractal geometry of surfaces in their relation to the microscopic roughness.

Appendix 1A   Viscoelasticity

For typical Newtonian fluids, the equation of motion is valid if the periods of the motion are larger than the relaxation times characterizing the molecules. However, this equation is not true for high viscous (non-Newtonian) fluids. Viscous fluids exist that behave as solids (for instance, glycerine and resin) during short intervals of time that are longer than the molecular relaxation times. Amorphous solids (for instance, glass) may be regarded as the limiting case of such non-Newtonian fluids having a large viscosity.

Resin when struck sharply shattered like glass, but it can flow slowly if sufficient time is allowed. This phenomenon was explained originally by Maxwell, who proposed a model consisting of an elastic spring and a viscous dashpot connected in series. Assume the system is under pure shear. The constitutive relation between the applied stress tensor σ_{ij} and the strain tensor e_{ij} in the spring obeys the Hooke law, $\sigma_{ij} = 2Ge_{ij}$, while the relation, $\sigma_{ij} = 2\eta\dot{e}_{ij}$, holds for the dashpot. Here, G is the modulus of rigidity and η is the viscosity. The total stain tensor is the sum of those in these two elements. Therefore, it is easy to see

$$\frac{d\sigma_{ij}}{dt} + \frac{\sigma_{ij}}{\tau} = 2G\frac{de_{ij}}{dt}, \tag{1A-1}$$

where $\tau = \eta/G$ is called the Maxwellian relaxation time. In periodic motion, where σ_{ij} and e_{ij} depend on the time through a factor $\exp(i\omega t)$, we obtain from Eq. (1A-1) that

$$\sigma_{ij} = \frac{2G}{1 - i/\omega\tau}e_{ij}. \tag{1A-2}$$

For $\omega\tau \ll 1$, this formula gives $\sigma_{ij} = 2i\omega\tau Ge_{ij} = 2\eta\dot{e}_{ij}$ and the material behaves as a Newtonian fluid, whereas for $\omega\tau \gg 1$, we have $\sigma_{ij} = 2Ge_{ij}$ and the material will behave as an amorphous solid. Eq. (1A-1) is the simplest differential representation of viscoelasticity [see Eq. (8.80)].

The integral representation of the isotropic Maxwellian constitutive equation can be written as

$$\sigma_{ij} = -p\delta_{ij} + 2\int_{-\infty}^{t} \mu(t-s)\dot{e}_{ij}(s)\,ds, \tag{1A-3}$$

where p is the pressure and

$$\mu(t) = \frac{\eta}{\tau} \exp\left(-\frac{t}{\tau}\right) \qquad (1A\text{-}4)$$

may be interpreted as the memory function. In the limit of $\tau \to 0$, the above two equations reduce to $\sigma_{ij} = -p\delta_{ij} + 2\eta\dot{e}_{ij}(s)$, which is the relation for a Newtonian fluid. In this book, we shall look beyond the Maxwell model and investigate the relationships between the viscoelasticity and the microstructure of complex materials.

References

1. M. Doi and S. F. Edwards, *The Theory of Polymer Dynamics* (Clarendon, Oxford 1986).
2. R. B. Bird, R. C. Armstrong, and O. Hassager, *Dynamics of Polymeric Liquids*, Vol. 2, 2nd ed. (Wiley, New York, 1987).
3. N. Wax (ed.), *Selected Papers on Noise and Stochastic Processes* (Dover, New York, 1954).
4. R. Kubo, M. Toda, and N. Hashitsume, *Statistical Physics II* (Springer-Verlag, Berlin, 1985).
5. L. D. Landau and E. M. Lifshitz, *Statistical Physics* (Pergamon, Oxford, 1969).
6. S. K. Ma, *Statistical Mechanics* (World Scientific, Philadelphia, 1985).
7. P. M. Chaikin and T. C. Lubensky, *Principles of Condensed Matter Physics* (Cambridge University, New York, 1995).
8. L. D. Landau and E. M. Lifshitz, *Fluid Mechanics* (Addison-Wesley, Reading, MA, 1959).
9. B. B. Mandelbrot, *The Fractal Geometry of Nature* (Freeman, New York, 1982).
10. A. Bunde and S. Havlin (eds.), *Fractals and Disordered Systems* (Springer-Verlag, Berlin, 1991).
11. F. Mallamace (ed.), *First International Conference on Scaling Concepts and Complex Fluids*, IL NUOVO CIMENTO, **16**(7), 1994.
12. J. D. Ferry, *Viscoelastic Properties of Polymers*, 3rd ed. (Wiley, New York, 1980).
13. L. C. E. Struik, *Physical Aging in Amorphous Polymers and Other Materials* (Elsevier, Amsterdam, 1978).
14. T. H. Hill, *Statistical Mechanics* (Dover, New York, 1987).
15. A.-L. Barabasi and H. E. Stanley, *Fractal Concepts in Surface Growth* (Cambridge University, New York, 1995).

2

Brownian Motion

It may seem surprising to start our study with Brownian motion; however, it is a paradigm of dissipative and irreversible behavior of a wide variety of systems. The concepts and methods developed for Brownian motion are the most fundamental cornerstones supporting stochastic processes and nonequilibrium statistical mechanics. These cornerstones are essential to the goal of investigating the time-dependent nonequilibrium properties of complex materials. We shall begin with the Markovian process for random walk. It is followed by a phenomenological description of the Langevin equation that links to diffusion and to stochastic relations. A general fluctuation theory is introduced as an example to show how noise with long-range correlation can be determined. This theory is coupled with the presentation of a molecular theory in the derivation of the Fokker–Plank equation. Finally, the important memory effect in the non-Markovian method for dissipative systems is discussed.

2.1 Markovian Process

The theory of Brownian motion is closely related to the mathematical analysis of random walk. Consider the process of a random walk. The probability density that a Brownian particle is at the point $\vec{r}$ after N step is denoted by $W_N(\vec{r})$. The first of these probability distributions is

$$W_1(\vec{r}) = \frac{\delta(|\vec{r}| - b)}{4\pi b^2}, \tag{2.1}$$

where b is the length of each step and δ is the Dirac delta function. For $N > 1$, the probability $W_N(\vec{r})\,d\vec{r}$ that lies in the vicinity $d\vec{r}$ of the point $\vec{r}$ can be calculated from

$$W_N(\vec{r}) = \frac{1}{4\pi b^2} \int W_{N-1}(\vec{r}')\delta(|\vec{r} - \vec{r}'| - b)\,d\vec{r}', \qquad (2.2)$$

where $\delta(|\vec{r} - \vec{r}'| - b)\,dr'/4\pi b^2$ is the transition probability. The Fourier transform of Eq. (2.1) is

$$W_1(\vec{q}) = \int W_1(\vec{r}) \exp(-i\vec{q} \cdot \vec{r})\,d\vec{r} = \frac{\sin qb}{qb}, \qquad (2.3)$$

where $\vec{q}$ is the wave vector and $q = |\vec{q}|$. Because $W_N(\vec{r})$ is related to $W_1(\vec{r})$ by $N - 1$ fold convolution product, its Fourier transform $W_N(\vec{q})$ is the Nth power of $W_1(\vec{q})$. Thus,

$$W_N(\vec{q}) = \left(\frac{\sin qb}{qb}\right)^N, \qquad (2.4)$$

and its Fourier inversion is

$$W_N(\vec{r}) = \frac{1}{(2\pi)^3} \int \left(\frac{\sin qb}{qb}\right)^N \exp(i\vec{q} \cdot \vec{r})\,d\vec{q}. \qquad (2.5)$$

The probability functions have been analyzed so far in the exact form.

Important approximation for $N \gg 1$ and $qb \ll 1$ for Eq. (2.4) results in

$$W(\vec{q}) = \lim_{N \to \infty} W_N(\vec{q}) = \left(1 - \frac{1}{6}q^2b^2 + \cdots\right)^N = \exp\left(-\frac{Nq^2b^2}{6}\right). \qquad (2.6)$$

Substituting Eq. (2.6) into Eq. (2.5) gives

$$W(\vec{r}) = \frac{1}{\left(\frac{2}{3}\pi Nb^2\right)^{2/3}} \exp\left(-\frac{3r^2}{2Nb^2}\right). \qquad (2.7)$$

The above process is the Markovian process for random walk, and Eq. (2.7) has a Gaussian distribution with respect to the position. Let $W(\vec{r})\,d\vec{r}$ be the probability of finding the Brownian particle at a distance between r and $r + dr$ in the volume $4\pi r^2 dr$ of the spherical shell. The mean square distance is determined by

$$\langle r^2 \rangle = \int r^2 W(\vec{r})\,d\vec{r} = \frac{4\pi}{\left(\frac{2}{3}\pi Nb^2\right)^{2/3}} \int_0^\infty r^4 \exp\left(-\frac{3r^2}{2Nb^2}\right)dr = nb^2, \qquad (2.8)$$

which plays an important role in polymer physics.

An alternative method exists to find Eq. (2.7) that can be generalized to more complicated problems in the case of $N \gg 1$ and $qb \ll 1$. Expanding both sides of Eq. (2.2) in Taylor series in N and $\Delta \vec{r}' = \vec{r}' - \vec{r}$, we obtain

$$W_N(\vec{r}) = W_{N-1}(\vec{r}) + \frac{\partial W}{\partial N} + \cdots \qquad (2.9)$$

and

$$W_{N-1}(\vec{r}') = W_{N-1}(\vec{r}) + \Delta\vec{r}' \cdot \frac{\partial W}{\partial \vec{r}} + \frac{1}{2}(\Delta\vec{r}')^2 \cdot \frac{\partial^2 W}{\partial \vec{r} \partial \vec{r}} + \cdots, \quad (2.10)$$

where N is treated as a continuous variable. Substituting eqs. (2.9) and (2.10) into Eq. (2.2), we arrive at the diffusion equation

$$\frac{\partial W}{\partial N} = \frac{b^2}{6}\nabla^2 W + O(b^4), \quad (2.11)$$

where ∇^2 is the Laplacian operator. Eq. (2.7) is the solution of Eq. (2.11). In the limit of $N \to \infty$, one can introduce a continuous time variable t. From eqs. (2.8) and (2.11), we get the diffusion coefficient

$$D = \langle r^2 \rangle / 6t, \quad (2.12)$$

and Eq. (2.11) becomes

$$\frac{\partial W}{\partial t} = D\nabla^2 W. \quad (2.13)$$

Let us consider that initially, W is nonzero only at $\vec{r} = 0$. The solution of this diffusion equation is

$$W(\vec{r}, t) = \frac{1}{2\sqrt{\pi(Dt)^3}} \exp\left(-\frac{r^2}{4Dt}\right), \quad (2.14)$$

which displays the characteristic Gaussian spreading of a random walk process. In nonequilibrium statistical mechanics, eqs. (2.7), (2.8), (2.12), and (2.14) clearly reveal the unique role of Brownian motion, which links the process from dynamic equations to the stochastic process [1].

2.2 Langevin Equation

A Brownian particle is considered to be subject to a fluctuating force exerted by the surrounding medium in addition to a frictional force. The Langevin equation has been introduced to describe such a motion:

$$m\frac{du(t)}{dt} = -\zeta u(t) + \tilde{F}(t), \quad (2.15)$$

where m is the mass of the particle, $\vec{u}$ is the velocity, ζ is the frictional coefficient, and $\tilde{F}$ is the random force. The mean of $\tilde{F}(t)$ over an ensemble of particles in thermal equilibrium is zero: $\langle \tilde{F}(t) \rangle = 0$. The random force autocorrelation function is treated as white noise:

$$g(t) \equiv \langle \tilde{F}(0)\tilde{F}(t) \rangle = A\delta(t). \quad (2.16)$$

A more detailed discussion of the correlation functions will be given in Section 3.2. The constant A is determined by the requirement

$$\langle u^2 \rangle = kT/m, \tag{2.17}$$

where k is the Boltzmann constant and T is temperature. Eq. (2.17) gives $A = 2\varsigma kT$. The use of the frictional force ςu implies the motion is steady. For spherical particle of radius a, the frictional coefficient is given by the Stokes law, $\varsigma = 6\pi\eta a$, where η is the viscosity of the liquid. The Stokesian dynamics (see Appendix 2A) is valid only for a low Reynolds number, $\rho_o ua/\eta$, where ρ_o is the density of the liquid. Taking into account that u is of order $(kT/m)^{1/2}$ and η is 10^{-2} poise, it is verified that the Reynolds number at all temperatures is small, provided $a > 10^{-8}$ cm and the density of the particle is not too small.

The formal solution of Eq. (2.15) is

$$u(t) = \left[u(0) + \int_0^t \tilde{F}(s)\exp(\gamma s)\,ds \right]\exp(-\gamma t), \quad \text{with } \gamma = \varsigma/m. \tag{2.18}$$

By taking the mean of Eq. (2.18), the expectation of the velocity decays as

$$\langle u(t) \rangle = u(0)\exp(-\gamma t) \tag{2.19}$$

because of the friction. The variance around the expectation can also be calculated from Eq. (2.18) by using Eq. (2.16)

$$\langle [u(t) - u(0)e^{-\gamma t}]^2 \rangle = e^{-2\gamma t}\int_0^t\int_0^t e^{\gamma(s_1+s_2)}\langle \tilde{F}(s_1)\tilde{F}(s_2)\rangle\,ds_1 ds_2$$

$$= \frac{kT}{m}[1 - \exp(2\gamma t)], \tag{2.20}$$

which grows in time, as mentioned in Eq. (2.14), and approaches to the Maxwellian value kT/m at $t \to \infty$.

For the velocity autocorrelation function $\psi(t)$, one finds the expression

$$\psi(t) \equiv \langle u(0)u(t)\rangle = \psi(0)\exp(-\gamma t) = \frac{kT}{m}\exp(-\gamma t). \tag{2.21}$$

In the theory of Brownian motion, the fundamental concern has always been the calculation of the mean square value of the displacement of the particle. It is given by

$$\langle x(t)^2 \rangle = \int_0^t ds_1 \int_0^t ds_2\, \langle u(s_1)u(s_2)\rangle. \tag{2.22}$$

This equation is transformed into

$$\lim_{t \to \infty} \frac{\langle x^2 \rangle}{2t} = \int_0^\infty \langle u(0)u(s)\rangle\,ds. \tag{2.23}$$

From eqs. (2.21), (2.23), and (2.12), one finds the diffusion coefficient

$$D = \int_0^\infty \psi(t)\,dt = kT/\varsigma. \tag{2.24}$$

This equation is the famous Einstein relation. Eq. (2.24) relates the Brownian motion to the thermal fluctuations of molecules, and it has received experimental verification. It is the first example of the most fundamental cornerstones in nonequilibrium statistical mechanics, often called the fluctuation–dissipation theorem in the literature. This theorem plays an important role in the study of colloidal dispersions and polymers, as we shall see in the later chapters.

It is a serious shortcoming of the theory that it relies on the frictional force for steady motion. Boussinesq found the force on a sphere in nonsteady motion [2]

$$-F(t) = \varsigma u(t) + \frac{1}{2}m_o\frac{du}{dt} + \alpha\pi^{-1/2}\int_{-\infty}^t (t-s)^{-1/2}\dot{u}(s)\,ds, \tag{2.25}$$

because of the retardation in viscous resistance caused by the effect of inertia (see Appendix 2A) on the Brownian motion. Here, $m_o = (4/3)\pi a^3\rho_o$ is the liquid displaced by the particle and $\alpha = \varsigma a(\rho_o/\eta)^{1/2}$. When the retarded friction is taken into account, the Langevin equation should be generalized to

$$m^*\frac{du}{dt} + \varsigma u + \alpha\pi^{-1/2}\int_{-\infty}^t (t-s)^{-1/2}\dot{u}(s)\,ds = \tilde{F}(t), \tag{2.26}$$

where $m^* = m + \frac{1}{2}m_o$. Introducing the Laplace transform

$$\psi(p) = \int_0^\infty \psi(t)\exp(-pt)\,dt, \tag{2.27}$$

we obtain velocity autocorrelation function

$$\psi(p) = \frac{(m^*/m)kT}{\varsigma + \alpha p^{1/2} + m^* p}, \tag{2.28}$$

which leads to

$$D = \int_0^\infty \psi(t)\,dt = \psi(p)|_\infty^0 = \frac{m^*kT}{m\varsigma} \tag{2.29}$$

instead of kT/ς. This process is not in accordance with the experimental fact that D/kT is equal to the mobility observed in the steady-state motion under the influence of a constant external force. We believe that this discrepancy is caused by an improper application of the Langevin equation. We shall discuss an alternative theory of Brownian motion in the next two sections on the basis of hydrodynamic fluctuation theory and nonequilibrium molecular theory. They will provide solid foundations in the derivation of correlation functions that are in complete agreement with Eq. (2.24).

2.3   Random Force Correlation

On the basis of the theory of fluctuations in fluid dynamics, the fluctuation–dissipation theorem is going to be derived for the Brownian motion in incompressible fluid. We shall relate the autocorrelation function $g(t)$ for random forces acting on a Brownian particle to the frictional coefficient, which determines the systematic force on the particle. The equation of motion and the continuity equation for an incompressible fluid (see Appendix 2A) are

$$\rho_o \frac{\partial w_i}{\partial t} = \tau_{ij,j} + s_{ij,j}, \qquad w_{i,i} = 0. \tag{2.30}$$

Here, w_i is the ith component of the velocity, τ_{ij} is the stress tensor, and s_{ij} is the stress tensor caused by spontaneous fluctuations in the liquid [3]. We use the subscript "j" to denote the differentiation with respect to x_i, and we use summation convention: A term in which a subscript occurs twice is a sum over that subscription. The s_{ij} has the following properties:

$$\langle s_{ij}(\vec{r}_1, t_1) s_{km}(\vec{r}_2, t_2) \rangle = 2kT C_{ijkm} \delta(\vec{r}_1 - \vec{r}_2) \delta(t_1 - t_2), \tag{2.31}$$

where

$$C_{ijkm} = \eta \left(\delta_{ik}\delta_{jm} + \delta_{im}\delta_{jk} - \frac{2}{3}\delta_{ij}\delta_{km} \right) \tag{2.32}$$

for isotropic Newtonian liquids.

It is convenient to introduce the Fourier time transform

$$g(\omega) = \int_{-\infty}^{\infty} g(t) \exp(i\omega t) \, dt, \tag{2.33}$$

which converts Eq. (2.30) into

$$\tau_{ij,j} + i\omega \rho_o w_i = -s_{ij,j}, \qquad w_{i,i} = 0. \tag{2.34}$$

We write the solution in the form

$$w_i = v_i + \tilde{v}_i, \qquad \tau_{ij} = \sigma_{ij} + \tilde{\sigma}_{ij}, \tag{2.35}$$

where $\tilde{v}_i$ and $\tilde{\sigma}_{ij}$ are fluctuation quantities and σ_{ij} and v_i are average quantities. The constitutive equation that relates σ_{ij} to the strain rate tensors can be written as

$$\tau_{ij} = -p\delta_{ij} + 2\eta \dot{e}_{ij}, \tag{2.36}$$

where p is the pressure and $\dot{e}_{ij} = (v_{i,j} + v_{j,i})/2$ is the rate of strain tensor. The equations to be solved are eqs. (2.37) and (2.38):

$$\sigma_{ij,j} + i\omega \rho_o v_i = 0, \qquad v_{i,i} = 0, \tag{2.37}$$

with the boundary conditions that $\vec{v} = \vec{u}(t)$, the particle velocity on the surface of the particle, and $\vec{v} = 0$ at infinity. Secondly,

$$\tilde{\sigma}_{ij,j} + i\omega\rho_o\tilde{v}_i = -s_{ij,j}, \qquad \tilde{v}_{i,i} = 0, \qquad (2.38)$$

with the condition that on the surface of the particle $\tilde{v}_i = 0$. The solution of Eq. (2.37) is the Boussinesq solution, which is mentioned earlier in Eq. (2.25). Eq. (2.38) will be used to calculate the random force $\tilde{F}_i$ acting on the particle [4,5].

We integrate the difference between the scalar products $v_i\tilde{\sigma}_{ij,j}$ and $\tilde{v}_i\sigma_{ij,j}$ over the volume of the liquid that surrounds the particle. By partial integration, we find

$$\int_V (v_i\tilde{\sigma}_{ij,j} - \tilde{v}_i\sigma_{ij,j})\,dV = -\int_S (v_i\tilde{\sigma}_{ij} - \tilde{v}_i\sigma_{ij})n_j\,dS, \qquad (2.39)$$

where n_j is the outward normal of the particle surface. On the right-hand side of the above equation, we use the fact that on the surface of the particle $v_i = u_i(t)$ and $\tilde{v}_i = 0$. Moreover, the integral of $\tilde{\sigma}_{ij}n_j$ over the particle surface is nothing but the random force $\tilde{F}_i$. On the left-hand side of Eq. (2.39) we use eqs. (2.37) and (2.38). The result is

$$u_i(\omega)\tilde{F}_i(\omega) = \int_V v_i s_{ij,j}\,dV. \qquad (2.40)$$

Using eqs. (2.31) and (2.32), we find

$$u_i(\omega_1)u_k(\omega_2)\langle\tilde{F}_i(\omega_1)\tilde{F}_k(\omega_2)\rangle = (kT/\pi)\delta(\omega_1 + \omega_2)$$
$$\times\left[-u_i(\omega_1)F_i(\omega_2) + i\omega_2\rho_o\int_V v_i(\omega_1)v_i(\omega_2)\,dV\right]. \qquad (2.41)$$

It is sufficient to apply this result to the one-dimensional model in which u, F, and $\tilde{F}$ are all in the same direction in t space,

$$\frac{\pi}{kT}\langle\tilde{F}(0)\tilde{F}(t)\rangle = \int_{-\infty}^{\infty} d\omega \exp(-i\omega t)\left[-\frac{F(\omega)}{u(\omega)} + i\omega\rho_o\int_V \frac{v_i(\omega)v_i^*(\omega)}{u_i(\omega)u_i^*(\omega)}\,dV\right], \qquad (2.42)$$

where the asterisk denotes the complex conjugated quantities. The ratio of F/u in the above equation is determined by Eq. (2.25). The Fourier transform of this equation yields

$$-F(\omega)/u(\omega) = \varsigma - \tfrac{1}{2}m_o i\omega - i\omega(1+i)\alpha(2\omega)^{-1/2} \equiv B(\omega). \qquad (2.43)$$

Making use of eqs. (2.37) and (2.38), the second term on the right-hand side under the volume integral in Eq. (2.42) can be reduced to

$$i\omega\rho_o\int_V v_i v_i^*\,dV = -\int_V v_i^*\sigma_{ij,j}\,dV = \int_S v_i^*\sigma_{ij}n_j\,dS + \int_V v_{i,j}^*\sigma_{ij}\,dV$$
$$= u^*F + 2\eta\int_V \dot{e}_{ij}^*\dot{e}_{ij}\,dV = iuu^*\,\text{Im}(F/u). \qquad (2.44)$$

Substituting eqs. (2.43) and (2.44) into Eq. (2.42), we obtain

$$g(\omega) \equiv \langle \tilde{F}(0)\tilde{F}(\omega) \rangle = 2kT\,\mathrm{Re}\,B(\omega), \tag{2.45}$$

which is another example of a more general principle called the fluctuation–dissipation theorem. This theorem provides a mechanism of relating energy dissipation to fluctuations in thermal equilibrium. Eq. (2.45) is a relationship connecting the correlation between fluctuations that occur spontaneously at different frequencies in equilibrium to the dynamic frictional coefficient. The explicit expression of Eq. (2.45) in time space is

$$\frac{\langle \tilde{F}(0)\tilde{F}(t) \rangle}{kT} = 2\varsigma\delta(t) - \frac{\alpha}{2\pi^{1/2}}t^{-3/2}. \tag{2.46}$$

This equation has a long-range correlation that differs from Eq. (2.16). The second term on the right-hand side of Eq. (2.46) is caused by the inertia of the liquid surrounding the Brownian particle. A retarded field is created that tends to take the opposite sign for the frictional force at $t > 0$.

2.4 Fokker–Planck Equation

The Brownian motion has been clearly described by the Langevin equation; however, it is phenomenological. We would like to consider a molecular theory that enables us to gain additional insight to the problem by looking at the simplest case of the translational movement in a viscous inert fluid. The molecular theory starts from the Hamiltonian for a system containing one heavy Brownian particle of mass m in a homogeneous solvent of N identical light particles of mass m_s. The time-dependent distribution function for the Brownian particle is defined as

$$f_B(\{1\}, t) = \int f(\{N\}, \{1\}, t)\,d\{N\}, \tag{2.47}$$

where $f(\{N\}, \{1\}, t)$ is the total distribution function of the system. Here $\{N\} = (\vec{p}^N, \vec{r}^N)$ denotes the momenta $\vec{p}^1, \ldots, \vec{p}^N$ and positions $\vec{r}^1, \ldots, \vec{r}^N$ of solvent particles; $\{1\} = (\vec{P}, \vec{R})$ represents momentum $\vec{P} = m\vec{u}$ and coordinates $\vec{R}$ of the Brownian particle. The Hamiltonian of the system is

$$H = H_B + H_o, \tag{2.48}$$

where

$$H_B = P_i P_i / 2m, \qquad H_o = p_i^N p_i^N / m_s + V(\vec{r}^N, \vec{R}),$$

V is the potential energy of solvent to solvent and solvent with the Brownian particle.

The total distribution function satisfies the Liouville equation (see Appendix 2B)

$$i\frac{\partial f}{\partial t} = (L_o + L_B)f, \tag{2.49}$$

which is an expression for the conservation of the distribution function. Here

$$i L_o = \frac{p_i^N}{m_s} \frac{\partial}{\partial \vec{r}_i^N} - \frac{\partial V}{\partial r_i^N} \frac{\partial}{\partial p_i^N},$$

$$i L_B = \frac{P_i}{m} \frac{\partial}{\partial R_i^N} + \tilde{F}_i \frac{\partial}{\partial P_i}.$$

$\tilde{F}_i = -\partial V/\partial R_i$ is the random force acting on the Brownian particle. Integrating Eq. (2.49) over $\{N\}$, we get

$$\frac{\partial f_B}{\partial t} + \frac{P_i}{m} \frac{\partial f_B}{\partial R_i} = -\int \tilde{F}_i \frac{\partial f}{\partial P_i} d\{N\}. \tag{2.50}$$

This equation is actually the generalization of the well-known BBGKY hierarchy [6].

The right-hand-side of Eq. (2.50) involves the total distribution function for which the formal solution is

$$f(t) = f(0) \exp(-i L_o t) \int_0^t \exp[-i(t - s)L_o] i L_B f(s) \, ds. \tag{2.51}$$

In order to compute $i L_B f$, we make the assumption

$$f = f_o f_B, \tag{2.52}$$

where f_o is the equilibrium distribution function of N identical solvent particles in the presence of a Brownian particle,

$$f_o = \exp(-H_o/kT) \Big/ \int \exp(-H_o/kT) \, d\{N\}. \tag{2.53}$$

Thus,

$$i L_B f(s) = [i L_B f_B + (P_k \tilde{F}_k/mkT) f_B] f_o. \tag{2.54}$$

To obtain the above equation, we have used the fact that

$$\langle \tilde{F}_i \rangle \equiv \int \tilde{F}_i f_o \, d\{N\} = 0, \tag{2.55}$$

as already mentioned in Eq. (2.15). Substituting Eq. (2.54) into Eq. (2.51), then into the right-hand side of Eq. (2.50), and using the relation $g(t) = \exp(i L_o t) g$ that displaces a function in time along a trajectory in phase space, Eq. (2.50) becomes

$$\frac{\partial f_B}{\partial t} + \frac{P_i}{m} \frac{\partial f_B}{\partial R_i} = \int_0^t \langle \tilde{F}_i \tilde{F}_k(t - s) \rangle \frac{\partial f}{\partial P_i} \left(\frac{\partial}{\partial P_k} + \frac{P_k}{mkT} \right) f_B(s) \, ds. \tag{2.56}$$

In most applications, the fluctuating forces are uncorrelated and the time correlation of the random force $g(t)$ is a white noise that has a δ-function decay [see Eq. (2.16)].

One then expects that Eq. (2.56) is reduced to the standard form of the Fokker–Planck equation.

When $g(t)$ has a long tail, as shown in Eq. (2.46), Eq. (2.56) has to be used in the calculation of the velocity correlation function. By defining

$$\psi(t) \equiv \langle u(o)u(t) \rangle_B = \int u(o)u(t) f_B \, d\{1\}, \qquad (2.57)$$

we can derive the relation between the velocity correlation function $\psi(t)$ and the random force correlation function $g(t)$:

$$\frac{d\psi(t)}{dt} = -\frac{1}{mkT} \int_o^t g(t-s)\psi(s) \, ds. \qquad (2.58)$$

Taking the Laplace transform defined by Eq. (2.27) and using Eq. (2.46), we have

$$\psi(p) = \frac{m\psi(0)}{\varsigma + \alpha p^{1/2} + mp}, \qquad (2.59)$$

where $\psi(o) = kT/m$. For the diffusion coefficient, this equation gives the correct answer $D = kT/\varsigma$ that is in contrast to Eq. (2.29). To find $\psi(t)$, we write $\varepsilon^2 = \alpha^2/4\xi m = 9\rho_o/8\rho$, $p = \gamma q$, and $x = \gamma t$. The Laplace inversion of Eq. (2.59) becomes

$$\frac{\psi(x)}{\psi(0)} = \frac{2\varepsilon}{\pi} \int_0^\infty \frac{s^{1/2} \exp(-xs)}{(s-1)^2 + 4\varepsilon^2 s} \, ds. \qquad (2.60)$$

This equation leads to the classical Stokes–Einstein decay, Eq. (2.21) for $\varepsilon = 0$. The asymptotic expression of Eq. (2.60) at large times is

$$\psi(t) = (\varepsilon/\pi^{1/2})\psi(0)(\gamma t)^{-3/2} + O(t^{-5/2}). \qquad (2.61)$$

This equation reveals that the random force with long-range correlation results in slow power law decay for the velocity correlation function at larger values of t. The slow power law decay is not limited to the inertial effect [7,8,9], but it is general phenomena related to the memory effect, which will be discussed in the next section.

2.5 Memory Effect

The physical quantities $\vec{x} = (x_a, x_b, \ldots)$ that describe a macroscopic body in equilibrium are close to their mean values. Nevertheless, fluctuations from the mean values become important in the study of nonequilibrium systems. The problem develops of finding transport coefficients from the probability distribution of these fluctuations. The probability for $\vec{x}$ to have values in the interval between $\vec{x}$ and $\vec{x} + d\vec{x}$ is proportional to $\exp[S(\vec{x})]$, where S is the entropy related to the irreversible process. By following the procedures developed by Landau and Lifshitz [10], the linear transport equation is written in the form

$$\dot{x}_a(t) = -\int_{-\infty}^{t} \lambda_{ab}(t-s)x_b(s)\,ds + y_a(t),\qquad(2.62)$$

where the summation convention for repeated indexes is used. The quantities $y_a(t)$ are random quantities and

$$\lambda_{ab}(t) = \lambda_{ba}(t)\qquad(2.63)$$

is the Onsager reciprocity relations. Eq. (2.62) has the form of a generalized Langevin equation mentioned in Section 2.2. As pointed out in Section 2.3, the random quantities are in general having long-range correlation. That is

$$\langle y_a(\omega_1)y_b(\omega_2)\rangle = 4\pi\langle x_a x_b\rangle\delta(\omega_1+\omega_2)\,\mathrm{Re}\,\lambda_{ab}[\omega_2],\qquad(2.64)$$

where $y(\omega)$ is the Fourier transform of $y(t)$ and $\lambda[\omega]$ is the Fourier–Laplace transform of $\lambda(t)$:

$$\lambda[\omega] = \int_0^{\infty} \lambda(t)\exp(i\omega t)\,dt.\qquad(2.65)$$

The average $\langle x_a x_b\rangle = \langle x_a(t)x_b(t)\rangle$ is determined by the matrix of the second derivative of the entropy [10,11,12], i.e.,

$$S = S_o - \tfrac{1}{2}\beta_{ab}x_a x_b,\qquad(2.66)$$

$$\langle x_a x_b\rangle = k\beta_{ab}^{-1}.\qquad(2.67)$$

It is convenient to rewrite these results in a slightly different form. The thermodynamic force can be written in the form

$$X_a = -\partial S/\partial x_a = \beta_{ab}x_b.\qquad(2.68)$$

If the quantities Λ_{ab} are defined by the equations

$$\lambda_{ab} = \Lambda_{ac}\beta_{cb},\qquad(2.69)$$

we can rewrite Eq. (2.62) as

$$\dot{x}_a(t) = -\int_{-\infty}^{t}\Lambda_{ab}(t-s)X_b(s)\,ds + y_a(t),\qquad(2.70)$$

and Eq. (2.64) becomes

$$\langle y_a(\omega_1)y_b(\omega_2)\rangle = 4\pi k\delta(\omega_1+\omega_2)\,\mathrm{Re}\,\Lambda_{ab}[\omega_2].\qquad(2.71)$$

A direct application of these general expressions to the problem of fluctuations in linear viscoelastic fluids should serve as an interesting example. For an incompressible fluid, the local stresses, rate of strains, and fluctuating stresses are related by

$$\sigma'_{ij}(t) = \int_{-\infty}^{t} C_{ijkm}(t-s)\dot{e}_{km}(s)\,ds + s_{ij}(t), \qquad (2.72)$$

where $\sigma'_{ij} = \sigma_{ij} + p\delta_{ij}$. This equation suggests that the random process is Markovian in the case of a Newtonian liquid; i.e., C_{ijkm} is a constant tensor and s_{ij} behaves like a white noise given by eqs. (2.32) and (2.31), respectively. For viscoelastic fluids, $C_{ijkm}(t)$ describe the extent to which the past flow influences the time-dependent behavior at the present and may be called the memory function. The random process given by Eq. (2.72) is not Markovian.

The rate of entropy production is determined by the rate of energy dissipation as

$$\dot{S} = \frac{1}{T} \int_{V} \sigma'_{ij}\dot{e}_{ij}\,dV. \qquad (2.73)$$

Comparing Eq. (2.70) with Eq. (2.72) and using Eq. (2.73), we let $\dot{x}_a$ correspond to σ'_{ij}, y_a to s_{ij}. Then, X_b correspond to $-\dot{e}_{km}/T$ and Λ_{ab} to TC_{ijkm}. This process gives the following result between random stress components:

$$\langle s_{ij}(\vec{r}_1, \omega_1)s_{km}(\vec{r}_2, \omega_2)\rangle = 4\pi kT\,\mathrm{Re}\,C_{ijkl}[\omega_2]\delta(\vec{r}_1 - \vec{r}_2)\delta(\omega_1 + \omega_2). \quad (2.74)$$

This equation is a generalization of Eq. (2.31). The right-hand side of Eq. (2.74) is not a delta function of time (or frequency) caused by the memory effect of non-Newtonian fluids, but it still has a delta function of space. This equation follows the constitutive relation between the local stress and local strain rate of linear viscoelastic fluids.

By way of illustration, we consider the Maxwell liquid (see Appendix 1A). The viscosity η in eqs. (2.32) and (2.72) for C_{ijkm} is replaced by the memory function

$$\mu(t) = (\eta/\tau)\exp(-t/\tau), \qquad (2.75)$$

which gives $\mu[\omega] = \eta/(1 - i\omega\tau)$, where τ is the relaxation time of the fluid. Replacing η in Eq. (2.43) by $\mu[\omega]$ and using Eq. (2.45), we get

$$\frac{\langle \tilde{F}(0)\tilde{F}(t)\rangle}{kT} = \frac{\varsigma}{\tau}\exp(-t/\tau) + \frac{\alpha}{\tau^{1/2}}\left\{2\delta(t) + \frac{d}{dt}[I_o(t/2\tau)\exp(-t/2\tau)]\right\}, \quad (2.76)$$

where I_o is the modified Bessel function of the first kind of order zero. In the limit $\tau \to 0$, the first term on the right-hand side of the above equation becomes $2\varsigma\delta(t)$ and the second term becomes $-\frac{1}{2}\alpha\pi^{-1/2}t^{-3/2}$, which is the result of a Newtonian fluid shown in Eq. (2.46). By substituting Eq. (2.76) into the integral equation (2.58), the velocity correlation function is obtained. The analytical expression is complex [13] and approaches to Eq. (2.60) as $\tau \to 0$. The numerical solution is shown in Figure 2.1, in which a comparison is made with the molecular dynamic calculation [14]. We clearly see that the viscoelastic relaxation has a significant influence on the dynamics and fluctuations of Brownian motion, and its influence on complex fluids and disordered solids will be a focus of this book.

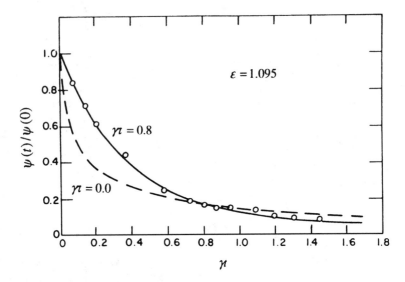

FIGURE 2.1. The velocity correlation function $\psi(t)$ is calculated as a function of time t, the viscoelastic parameter $\gamma\tau$, and the inertia parameter $\varepsilon = 9\rho_0/8\rho$. The relaxation time of the fluid is τ, γ is defined in Eq. 2.18, the density of the fluid is ρ_0, and the density of the Brownian particle is ρ. The dotted line is based on Eq. (2.60) and represents the inertia effect alone. The circles are the data from the molecular dynamic calculations [14].

Appendix 2A   The Navier–Stokes Equation

Three local conservation laws of the mass flux, the momentum flux, and the energy flux densities govern the motion of viscous fluid. For our purpose, only the former two densities will be written here. In addition, a constitutive equation is needed to relate the stress tensor σ_{ij} to the rate of strain tensor expressed by the velocity field. All physical quantities are functions of the coordinates x_i and of the time t.

The continuity equation is

$$\partial\rho/\partial t + div\,(\rho\vec{v}) = 0, \tag{2A-1}$$

where ρ is the fluid density and $\vec{v}$ is the velocity.

The equation of motion is

$$\rho\left(\frac{\partial v_i}{\partial t} + v_j\frac{\partial v_i}{\partial x_j}\right) = \frac{\partial\sigma_{ij}}{\partial x_j}, \tag{2A-2}$$

without including the body force density.

The constitutive equation is

$$\sigma_{ij} = \left(-p + \kappa\frac{\partial v_l}{\partial x_l}\right)\delta_{ij} + \eta\left(\frac{\partial v_i}{\partial x_j} + \frac{\partial v_j}{\partial x_i} - \frac{2}{3}\delta_{ij}\frac{\partial v_l}{\partial x_l}\right) \tag{2A-3}$$

when the fluid is isopropic and Newtonian. Here, p is the pressure, η is the shear viscosity, and κ is the dilatational viscosity, which is usually negligible.

Combining eqs. (2A-2) and (2A-3) gives

$$\rho\left[\frac{\partial \vec{v}}{\partial t} + (\vec{v} \cdot \mathbf{grad})\vec{v}\right] = -\mathbf{grad}\,p + \eta\nabla^2\vec{v} + (\kappa + \eta/3)\mathbf{grad}(div\,\vec{v}). \quad (2A\text{-}4)$$

If the fluid is incompressible, Eq. (2A-1) becomes $div\,\vec{v} = 0$ and the last term on the right-hand side of Eq. (2A-4) is zero. Thus, we get

$$\rho\left[\frac{\partial \vec{v}}{\partial t} + (\vec{v} \cdot \mathbf{grad})\vec{v}\right] = -\mathbf{grad}\,p + \eta\nabla^2\vec{v}, \quad (2A\text{-}5)$$

which is called the Navier–Stokes equation. On the left-hand side, the first term is the inertia term and the second is the convection term. Both of them are negligible in the case of low Reynolds number flow when the viscosity is high and the particle is small. These assumptions are behind Eq. (4.9) for the Stokesian dynamics in Chapter 4, which also leads to the familiar Stokes law of the friction coefficient used in the Langevin equation [see Eq. (2.15)]. The equation of motion for the Boussinesq flow [see Eq. (2.25)] follows directly from Eq. (2A-5) by keeping the inertia term but neglecting the convection term. Finally, Eq. (2A-3) has to be replaced by Eq. (1A-3) in the case of viscoelastic fluids (see Section 2.5).

Appendix 2B   The Liouville Theorem

Let us consider a macroscopic system of N particles that move in accordance with the Hamilton equations:

$$\dot{\vec{r}}_i = \frac{\partial H}{\partial \vec{p}_i}, \qquad \dot{\vec{p}}_i = -\frac{\partial H}{\partial \vec{r}_i}, \qquad i = 1, 2, \ldots, N, \quad (2B\text{-}1)$$

where $\vec{r}_i$ and $\vec{p}_i$ are the coordinates and the conjugate momenta of the ith particle and H is the Hamiltonian. The path of the phase point is determined by Eq. (2B-1). For very large N, a detailed microscopic description like this would have no practical value. Because we are interested in the macroscopic properties of the system, the average behavior of the particles is important. Therefore, we try to describe the system by a probability density $f(\vec{r}_1, \ldots, \vec{r}_N, \vec{p}_1, \ldots, \vec{p}_N, t)$.

The conservation law governs the time evolution of the density f, which is the exact $6N$-dimensional analog of the continuity equation for fluid mechanics. The mathematical expression for the conservation of the probability density is

$$\frac{df}{dt} = 0. \quad (2B\text{-}2)$$

It can be written more explicitly in the compact form:

$$\frac{df}{dt} + [f, H] = 0, \quad (2B\text{-}3)$$

where $[f, H]$ denotes the Poisson bracket

$$[f, H] = \sum_i \left(\frac{\partial f}{\partial \vec{r}_i} \frac{\partial H}{\partial \vec{p}_i} - \frac{\partial f}{\partial \vec{p}_i} \frac{\partial H}{\partial \vec{r}_i} \right). \tag{2B-4}$$

Eq. (2B-3) is called the Liouville theorem.

References

1. S. Chandrasekhar, Rev. Mod. Phys. **15**, 1 (1943).
2. J. Boussinesq, *Theorie Analytique de la Chaleur, II* (Gauthiers-Villars, Paris, 1903).
3. L. D. Landau and E. M. Lifshitz, *Fluid Mechanics* (Addison-Wesley, Reading, MA, 1959).
4. R. F. Fox and G. E. Uhlenbeck, Phys. Fluids **13**, 1893 (1970).
5. T. S. Chow and J. J. Hermans, J. Chem. Phys. **56**, 3150 (1972).
6. G. E. Uhlenbeck and G. W. Ford, *Lectures in Statistical Mechanics* (American Mathematical Society, Providence, RI, 1963).
7. M. H. Ernst, E. F. Hauge, and J. M. J. VanLeeuwen, Phys. Rev. Lett. **25**, 1254 (1970).
8. R. Zwanzig and M. Bixon, Phys. Rev. A **2**, 2005 (1970).
9. A. Widom, Phys. Rev. A **3**, 1374 (1971).
10. L. D. Landau and E. M. Lifshitz, *Statistical Physics* (Pergamon, Oxford, 1969).
11. S. R. De Groot, *Thermodynamics of Irreversible Processes* (North-Holland, Amsterdam, 1969).
12. L. Onsager and S. Machlup, Phys. Rev. **91**, 1505 (1953).
13. T. S. Chow, J. Chem. Phys. **61**, 2868 (1974).
14. G. Subramanian, D. Levitt, and H. T. Davis, J. Chem. Phys. **60**, 591 (1974).

3

Dynamic Response

In order to study the intrinsic properties of complex fluids and disordered solids in nonequilibrium situations, a general formulation is needed to describe the dynamic response of a system subjected to an external force field. We shall see in this chapter how the linear response theory relates the nonequilibrium properties of a system directly to fluctuations in equilibrium and, thus, has almost the same validity as equilibrium statistical mechanics. Three commonly used methods of analyzing the nonequilibrium properties of the system are response, relaxation, and susceptibility. Response measures the time evolution of the system under the influence of a time-independent force. Relaxation studies the time-dependent decay after a constant force is removed. Susceptibility investigates the frequency-dependent response to an oscillatory force. We shall see that these properties are not independent of each others. The linear response theory will be used to establish the relationships between the three intrinsic properties of materials. The master equation is fundamental in the study of nonequilibrium stochastic processes, and we shall see it often in the succeeding chapters as a good starting point for analyzing and interpreting cooperative and disordered systems that may be complex fluids, solids, or interfaces.

3.1 Linear Response Theory

An important approach to the study of the nonequilibrium properties of a many-body system is to calculate the response of the system to an applied perturbation. Let us consider a classical system. At the time $t \to -\infty$, the system is described by a Hamiltonian H_0, and at some later time, the total Hamiltonian is taken to be

$$H = H_0 + H_1 \equiv H_0 - A \cdot F(t), \tag{3.1}$$

where F is an external force and A is a "displacement." Both F and A may be scalars or vectors. The equation of motion for the total distribution function is determined by the Liouville equation

$$\frac{\partial f}{\partial t} + [f, H] = 0, \tag{3.2}$$

where $[f, H]$ represents the Poisson bracket [see Eq. (2B-4)].

We seek the solution

$$f = f_0 + f_1, \tag{3.3}$$

where f_0 satisfies $[f_0, H_0] = 0$ at $t \to -\infty$ when $F = 0$. Neglecting a second-order term, the linear differential equation for f_1 is

$$\frac{\partial f_1}{\partial t} + [f_1, H_0] = -[f_0, H_1]. \tag{3.4}$$

To solve the above equation, let us use the Green's function G satisfying

$$\frac{\partial G(t-s)}{\partial t} + [G(t-s), H_0] = \delta(t-s). \tag{3.5}$$

Formally, one gets

$$G(t-s) = \exp i L(t-s), \quad \text{for } t > s, \tag{3.6}$$

as we discussed in Section 2.4. Here, the Liouville operator is defined such that $iLG + [G, H_0] = 0$. Thus,

$$f_1(t) = -\int_{-\infty}^{t} [\exp i L(t-s)][f_0, H_1]_s \, ds. \tag{3.7}$$

The nonequilibrium average of a dynamic quantity B is obtained from

$$\langle B(t) \rangle_{noneq} = Tr\{Bf(t)\} \equiv \int Bf(t) \, d\Gamma, \tag{3.8}$$

where $d\Gamma = d\vec{p}d\vec{q}$ represents the volume elements of the phase space $(\vec{p}, \vec{q})$, and the distribution function is normalized to one. Substituting Eq. (3.3) into Eq. (3.8) yields $\langle B \rangle_{noneq} \equiv \langle B \rangle_{eq} + \Delta\langle B \rangle$. The equilibrium average is

$$\langle B \rangle_{eq} \equiv \langle B \rangle = \int Bf_0 \, d\Gamma, \tag{3.9}$$

and the contribution from the perturbation is

$$\Delta\langle B\rangle = \int B f_1(t)\, d\Gamma = -\int d\Gamma \int_{-\infty}^{t} \{\exp[iL(t-s)][f_0, A]_s\} \cdot F(s) B\, ds$$

$$= -\int d\Gamma \int_{-\infty}^{t} [f_0, A]_s B(t-s) \cdot F(s)\, ds, \qquad (3.10)$$

where Eq. (3.7) and the relation $B(t) = \exp(iLt)B$ were used.

A linear response function [1,2] can now be introduced as

$$\chi_{AB}(t-s) = \int d\Gamma [A, f_0]_s B(t), \qquad (3.11)$$

where $[A, f_0] = -[f_0, A]$. Eq. (3.11) shows that χ_{BA} is an averaged commutator. Hence, Eq. (3.10) becomes

$$\Delta\langle B(t)\rangle = \int_{-\infty}^{t} \chi_{AB}(t-s) \cdot F(s)\, ds. \qquad (3.12)$$

This process is an important expression in the study of the nonequilibrium behavior of dissipated systems, and the response function may also be interpreted as the memory function related to the non-Markovian process mentioned in Section 2.5. In a viscoelastic medium, the response of the displacements lags behind the forces. The forces in the past enter, but those in the future do not because of causality. Accordingly, the present state reflects the memory of history.

3.2 Correlation Functions

The expression for χ_{AB} in Eq. (3.11) is general. We would like to understand the physics of the response function and to establish its connection to the relaxation function. When either of these functions is given, the other can be obtained. To analyze these functions quantitatively, we need to talk about their connection to the correlation of spontaneous fluctuations, which is done with the language of time correlation functions [2,3]. The fluctuation in the dynamic quantity $B(t)$ from its time-independent equilibrium average is $\Delta B(t) = B(t) - \langle B\rangle$, and the fluctuation in another dynamic quantity A is ΔA. The equilibrium correlation between ΔA and ΔB at different times is

$$C_{AB}(t_1, t_2) = \langle \Delta A(t_1)\Delta B(t_2)\rangle = \langle A(t_1)B(t_2)\rangle - \langle A(t_1)\rangle\langle B(t_2)\rangle. \qquad (3.13)$$

The correlation between dynamic variables at different times should depend on the difference between these times rather than on the absolute value of time. Because of the translational invariant in time ($t = t_2 - t_1$), Eq. (3.13) is usually written in the form

$$C_{AB}(t) = \langle \Delta A(0)\Delta B(t)\rangle = \langle A(0)B(t)\rangle - \langle A(0)\rangle\langle B(t)\rangle, \qquad (3.14)$$

where $t_1 = 0$ is assumed.

We would like to transform the general expression for χ_{AB} in Eq. (3.11) into a more convenient form. Let us look at the equilibrium distribution function mentioned in Eq. (3.3):

$$f_0 = \frac{\exp(-\beta H_0)}{Tr\{\exp(-\beta H_0)\}}, \qquad \beta = 1/kT. \tag{3.15}$$

By making use of this equation, the Poisson bracket shown in Eq. (3.11) can be written explicitly as

$$[A, f_0]_s \equiv [A(s), f_0] = \sum_i \left(\frac{\partial A(s)}{\partial p_i} \frac{\partial f_0}{\partial q_i} - \frac{\partial A(s)}{\partial q_i} \frac{\partial f_0}{\partial p_i} \right)$$

$$= \beta f_0 [H_0, A(s)] = \beta f_0 \dot{A}(s). \tag{3.16}$$

Substituting Eq. (3.16) into Eq. (3.11) and setting $s = 0$, we get

$$\chi_{AB}(t) = \beta \int d\Gamma f_0 \dot{A}(0) B(t) \equiv \beta \langle \dot{A}(0) B(t) \rangle$$

$$= -\beta \langle A(0) \dot{B}(t) \rangle = -\beta \frac{d}{dt} \langle A(0) B(t) \rangle \tag{3.17}$$

for $-\infty < t < \infty$. These principal results link the response function of a nonequilibrium system to the correlation of dynamic variables in the equilibrium state.

To understand the physical meaning of the response function, let us assume an impulse force acting to the system at $t = 0$ that is represented by a δ-function:

$$F(t) = F_\delta \delta(t), \tag{3.18}$$

where F_δ is the strength of the impulse force. Substituting Eq. (3.18) into Eq. (3.12) gives

$$\Delta \langle B(t) \rangle / F_\delta = \int_{-\infty}^t \chi_{AB}(t - s) \delta(s) \, ds = \chi_{BA}(t), \quad \text{for } t > 0. \tag{3.19}$$

The function χ_{AB} is expected to increase at short times and then vanish in a sufficiently long time when the system returns to thermal equilibrium. On the other hand, the relaxation phenomena is observed in the case of a sudden removal of a force F_0 kept constant from $t \to -\infty$ to $t = 0$:

$$F(t) = F_0 \theta(-t), \tag{3.20}$$

where θ is the unit step function. By using Eq. (3.20) and noting that $d\theta(-s)/dt = -\delta(t)$, the dynamic response given by Eq. (3.12) is reduced to

$$\Delta \langle B(t) \rangle / F_0 = \int_t^\infty \chi_{AB}(s) \, ds \equiv \Phi_{BA}(t). \tag{3.21}$$

The function Φ_{AB} is called the relaxation function [1,2]. The system starts from a equilibrium value at $t = 0$ and then decays toward another equilibrium value at large times. In addition, Eq. (3.21) defines the static susceptibility [4]: $\Delta\langle B(t=0)\rangle/F_0 = \Phi_{AB}(0)$. Substituting Eq. (3.17) into Eq. (3.21) yields

$$\Phi_{AB}(t) = -\int_t^\infty \frac{d}{ds}\langle A(0)B(s)\rangle\,ds$$

$$= \beta\left[\langle A(0)B(t)\rangle - \lim_{t\to\infty}\langle A(0)B(t)\rangle\right]. \qquad (3.22)$$

According to the law of increase in entropy, no correlation can exist between physical quantities that are well separated in time; i.e.,

$$\lim_{t\to\infty}\langle A(0)B(t)\rangle = \langle A(0)\rangle\langle B(t)\rangle. \qquad (3.23)$$

By using eqs. (3.14) and (3.23), Eq. (3.22) becomes

$$\Phi_{AB}(t) = \beta C_{AB}(t) = \beta[\langle A(0)B(t)\rangle - \langle A(0)\rangle\langle B(t)\rangle]. \qquad (3.24)$$

The second term on the right-hand side of the above equation guarantees that $\Phi_{AB}(t)$ and $C_{AB}(t)$ vanishes as $t\to\infty$. For $t=0$, an explicit expression for the static susceptibility

$$\Phi_{AB}(0) = \beta[\langle A(0)B(0)\rangle - \langle A\rangle\langle B\rangle] \qquad (3.25)$$

is obtained. Clearly, eqs. (3.14), (3.21), (3.22), and (3.24) give

$$\chi_{AB}(t) = -\frac{d\Phi_{AB}(t)}{dt}$$

$$= -\beta\frac{d\langle\Delta A(0)\Delta B(t)\rangle}{dt}, \quad \text{for } -\infty < t < \infty. \qquad (3.26)$$

This equation is an alternative form of Eq. (3.17). Eq. (3.26) is the sought after general formula that relates χ_{AB}, Φ_{AB}, and spontaneous fluctuations.

As an example, let us consider the simplest case of Brownian motion mentioned in Chapter 2. The Hamiltonian $H_1 = -\vec{r}\cdot\vec{F}$ and the force $\vec{F}$ is related to the velocity $\vec{u}$ by $\vec{F} = \varsigma\vec{u} = \varsigma\dot{\vec{r}}$, where ς is the frictional coefficient. When the dynamic quantities A and B in Eq. (3.17) are replaced by $\vec{r}$ and $\vec{u}$, respectively, the complex mobility of a Brownian particle is

$$\mu[\omega] = \beta\int_0^\infty\langle u(0)u(t)\rangle e^{i\omega t}\,dt = \frac{\beta}{3}\int_0^\infty\langle\vec{u}(0)\cdot\vec{u}(t)\rangle e^{i\omega t}\,dt, \qquad (3.27)$$

by taking the Fourier–Laplace transform of Eq. (3.17). Kubo [1,5] calls Eq. (3.27) the first fluctuation–dissipation theorem. It is a direct result of the linear response theory.

3.3 Generalized Susceptibility

We have just discussed the transient response to a time-dependent field, and we now turn our attention to the evaluation of the frequency-dependent response of a system to an oscillatory field. The complex response, characterized by a generalized susceptibility, is going to be given by the frequency Fourier transform of the fluctuations of the system in the absence of external field. The generalized susceptibility can be the permittivity in the dielectric case, or the compliance of viscoelastic liquid or solid. Let us return to Eq. (3.12) by changing the integration variable that yields

$$\Delta \langle B(t) \rangle = \int_0^\infty \chi_{AB}(s) \cdot F(t - s) \, ds. \tag{3.28}$$

The disturbance at time s can lead to change in $\langle B(t) \rangle$ only for times $t > s > 0$ that means the response of $\langle B(t) \rangle$ to $F(t - s)$ is causal. When Eq. (3.28) is subject to the Fourier transform in time [see Eq. (2.33)], the relation between $\langle B(\omega) \rangle$ to $F(\omega)$ is

$$\Delta \langle B(\omega) \rangle = \chi[\omega] \cdot F(\omega), \tag{3.29}$$

where ω is the angular frequency and the subscripts were dropped temporary for simplicity.

The Fourier–Laplace transform of the response function is defined as a generalized susceptibility

$$\chi[\omega] = \int_0^\infty \chi(t) \exp(i\omega t) \, dt, \tag{3.30}$$

which depends on the properties of materials. The zero-frequency limit of Eq. (3.30) is equal to the zero-time limit of Eq. (3.21); i.e.

$$\chi[\omega = 0] = \int_0^\infty \chi(t) \, dt = \Phi(t = 0), \tag{3.31}$$

which gives the static susceptibility [see Eq. (3.25)].

The function $\chi[\omega]$ is complex, and let us split it into the real and imaginary parts as

$$\chi[\omega] = \chi'(\omega) + i\chi''(\omega). \tag{3.32}$$

The definition (3.30) shows immediately that

$$\chi[-\omega] = \chi^*[\omega], \tag{3.33}$$

as the requirement of causality. Separating the real and imaginary parts, we find

$$\chi'(-\omega) = \chi'(\omega), \qquad \chi''(-\omega) = -\chi''(\omega); \tag{3.34}$$

i.e., $\chi'(\omega)$ is an even function of frequency, and $\chi''(\omega)$ an odd function.

In general, any analytical function $\chi(z)$ of a complex variable z can be expressed by Cauchy's integral as

$$\chi(z) = \frac{1}{\pi i} P \int_{-\infty}^\infty \frac{\chi(\omega)}{\omega - z} \, d\omega, \tag{3.35}$$

where z is in the upper half of complex ω-plane and the symbol P denotes the principal value integral. This equation is again a mathematical expression as a result of causality. If χ does not have singularity in the upper half of a ω-plane, eqs. (3.32) and (3.35) give

$$\chi'(z) = \frac{1}{\pi} P \int_{-\infty}^{\infty} \frac{\chi''(\omega)}{\omega - z} \, d\omega \qquad (3.36)$$

and

$$\chi''(z) = -\frac{1}{\pi} P \int_{-\infty}^{\infty} \frac{\chi'(\omega)}{\omega - z} \, d\omega. \qquad (3.37)$$

Each of these two formulae are equivalent to Eq. (3.35) because they are each other's Hilbert transforms. These dispersion relations are known as the Kramers–Kronig relations that relate the real and imaginary parts of $\chi[\omega]$ to each other. With the properties given by eqs. (3.33) and (3.34), we get an alternative form of the Kramers–Kronig relations:

$$\chi'(\omega) = \frac{2}{\pi} \int_{0}^{\infty} \frac{\chi''(\xi)\xi}{\xi^2 - \omega^2} \, d\xi + const \qquad (3.38)$$

and

$$\chi''(\omega) = -\frac{2\omega}{\pi} \int_{0}^{\infty} \frac{\chi'(\xi)}{\xi^2 - \omega^2} \, d\xi. \qquad (3.39)$$

These dispersion relations have a very wide validity. They enable the imaginary part of $\chi[\omega]$ to be determined by the known real part, or vice versa. They are also used as a check on experimental results when both the real and imaginary parts of the generalized susceptibility are measured.

3.4   Fluctuation–Dissipation Theorem

The essence of this theorem is the focus of this section because it serves as the foundation of irreversible statistical mechanics. We shall see how the generalized susceptibility is related (1) to the energy dissipation of a dynamic system via the linear response theory and (2) to the intrinsic fluctuations in a thermal equilibrium system via frequency correlation functions.

The function $\chi''(\omega)$ plays a important role of irreversibility. We shall prove that $\chi''(\omega)$ is related to the dissipation of a system to an oscillatory field. The determination of the susceptibility often involves the measurement of power absorbed. In order to see how this relationship emerges from a general consideration, we recall that the perturbation Hamiltonian [see Eq. (3.1)] has the form

$$H_1 = -\sum_i A_i F_i(t). \qquad (3.40)$$

The rate of work is determined from

$$\frac{dW}{dt} = -\sum_i \Delta\langle A_i(t)\rangle \dot{F}_i(t)$$

$$= -\sum_{ij} \int_{-\infty}^{t} \dot{F}_i(t)\chi_{ij}(t-s)F_j(s)\,ds. \tag{3.41}$$

In the presence of an oscillatory force, we have

$$F_i(t) = \tfrac{1}{2}[F_i \exp(-i\omega t) + F_i^* \exp(i\omega t)]. \tag{3.42}$$

Now, what is actually measured in an experiment is the average power absorbed over a cycle. The average rate of work is

$$\frac{\overline{dW}}{dt} = -\frac{i\omega}{4}\sum_{ij}(F_i^*\chi_{ij}[\omega]F_j - F_i\chi_{ij}[-\omega]F_j^*). \tag{3.43}$$

By using the properties of eqs. (3.32) and (3.33), Eq. (3.43) becomes

$$\frac{\overline{dW}}{dt} = \frac{\omega}{2}\sum_{ij}[F_i^*\chi_{ij}''(\omega)F_j] > 0, \tag{3.44}$$

which shows that $\omega\chi''(\omega) > 0$ as the cause of the energy dissipation in a stable dynamic system.

The properties of the real and imaginary parts of the generalized susceptibility have already been discussed. The next step is to see how the generalized susceptibility is expressed by frequency correlation functions of the dynamic quantities A and B from their equilibrium averages. The Fourier transform of a time correlation function $C_{AB}(t)$ defined by Eq. (3.14) is

$$C_{AB}(\omega) = \int_{-\infty}^{\infty} C_{AB}(t)\exp(i\omega t)\,dt$$

$$\equiv \int_{-\infty}^{\infty}[\langle \Delta A(0)\Delta B(t)\rangle]\exp(i\omega t)\,dt, \tag{3.45}$$

In this form, it is clear that $C_{AB}(\omega)$ is the spectrum of spontaneous fluctuations. From Eq. (3.26), the correlation function $C_{AB}(t)$ is related to the response function $\chi_{AB}(t)$ by

$$\chi_{AB}(t) = -\beta\frac{dC_{AB}(t)}{dt}, \quad \text{for } -\infty < t < \infty. \tag{3.46}$$

Because $C_{AB}(t)$ is an even function, the above equation suggests that $\chi_{AB}(t)$ is an odd function.

The Fourier transform of the right-hand side of Eq. (3.46) is

$$-\beta\int_{-\infty}^{\infty}\frac{dC_{AB}(t)}{dt}\exp(i\omega t)\,dt = i\beta\omega C_{AB}(\omega), \tag{3.47}$$

and the left-hand side is

$$\int_{-\infty}^{0} \chi_{AB}(t) \exp(i\omega t)\, dt + \int_{0}^{\infty} \chi_{AB}(t) \exp(i\omega t)\, dt$$

$$= -\chi_{AB}[-\omega] + \chi_{AB}[\omega] = -\chi_{AB}^{*}[\omega] + \chi_{AB}[\omega] = 2i\,\chi_{AB}''(\omega), \qquad (3.48)$$

where eqs. (3.32) and (3.33) were used. Hence,

$$\chi_{AB}''(\omega) = \frac{\beta\omega}{2} C_{AB}(\omega) \equiv \frac{\beta\omega}{2} \langle \Delta A(0) \Delta B(\omega) \rangle. \qquad (3.49)$$

This relationship between dissipation and correlation spectrum is commonly called the fluctuation–dissipation theorem [1,5]. Eq. (3.48) can also be written in the form

$$\chi_{AB}(\omega) = \int_{-\infty}^{\infty} \chi_{AB}(t) \exp(i\omega t)\, dt$$

$$= 2i\,\chi_{AB}''(\omega) = \int_{-\infty}^{\infty} 2i\,\chi_{AB}''(t)\theta(t)\exp(i\omega t)\, dt. \qquad (3.50)$$

where $\theta(t)$ is the unit step function. Therefore,

$$\chi_{AB}(t) = 2i\,\chi_{AB}''(t)\theta(t). \qquad (3.51)$$

This process is an interesting expression for the response function that reveals its salient feature of dissipation and causality. The equivalence between the response function, generalized susceptibility, and correlation function has now been clearly shown.

3.5   Non-Markovian and Nonlocal Relations

In the case of nonuniform systems, the force and displacement mentioned in Section 3.1 should include the dependence of the spatial coordinate $(\vec{r})$ in a general formula for the linear response theory. This section also summarizes the important results that link several dynamic quantities and functional relations. We write the part of the Hamiltonian containing the applied field as

$$H_1 = -\sum_i \int d\vec{r}\, A_i(\vec{r}) F_i(\vec{r}, t), \qquad (3.52)$$

and it is our aim to determine the response in the form [see Eq. (3.12)]

$$\Delta\langle B_j(\vec{r}, t)\rangle = \sum_i \int d\vec{r}' \int_{-\infty}^{t} \chi_{ij}(\vec{r}, \vec{r}', t - s) F_i(\vec{r}', s)\, ds. \qquad (3.53)$$

The response function is given by

$$\chi_{ij}(\vec{r}, \vec{r}', t) = \langle [A_i(\vec{r}'), B_j(\vec{r}, t)] \rangle, \qquad (3.54)$$

with the thermal average carried out over the unperturbed system:

$$\langle A_i \rangle = Tr\{f_0 A_i\}. \tag{3.55}$$

Besides the non-Markovian property [see Section 2.5], the nonlocal properties at different points in a system are taken into account. As in the case of time, the invariant under the space translation is assumed so that the response and other functions depend only on the difference of position vectors $\vec{r} - \vec{r}'$. Thus,

$$\chi_{ij}(\vec{r}, \vec{r}', t) = \chi_{ij}(\vec{r} - \vec{r}', t) = \chi_{ij}(\vec{r}, t) \tag{3.56}$$

by setting $\vec{r}' = 0$. The relaxation function is related to the response function by the generalization of Eq. (3.21)

$$\Phi_{ij}(\vec{r}, t) = \int_t^\infty \chi_{ij}(\vec{r}, s)\, ds, \tag{3.57}$$

which satisfies

$$\chi_{ij}(\vec{r}, t) = -\frac{\partial \Phi_{ij}(\vec{r}, t)}{\partial t}, \quad \text{for } -\infty < t < \infty. \tag{3.58}$$

If we introduce the Fourier transform in both time and space,

$$B_j(\vec{q}, \omega) = \int_{-\infty}^\infty dt \int d\vec{r}\, B_j(\vec{r}, t) \exp(i\omega t - i\vec{q} \cdot \vec{r}). \tag{3.59}$$

Eq. (3.53) can then be written in the form

$$\Delta \langle B_j(\vec{q}, \omega) \rangle = \sum_i \chi_{ij}[\vec{q}, \omega] F_i(\vec{q}, \omega). \tag{3.60}$$

Here, the generalized susceptibility follows the Fourier transform in space, but the Fourier–Laplace transform in time

$$\chi_{ij}[\vec{q}, \omega] = \int_0^\infty dt \int d\vec{r}\, \chi_{ij}(\vec{r}, t) \exp(i\omega t - i\vec{q} \cdot \vec{r})$$
$$= \chi'_{ij}(\vec{q}, \omega) + i\chi''_{ij}(\vec{q}, \omega), \tag{3.61}$$

In terms of dissipation function, the above equation can be written in the following form by using Eq. (3.36):

$$\chi_{ij}[\vec{q}, \omega] = \frac{1}{\pi} P \int_{-\infty}^\infty \frac{\chi''_{ij}(\vec{q}, \xi)}{\omega - \xi}\, d\xi + i\chi''_{ij}(\vec{q}, \omega). \tag{3.62}$$

From eqs. (3.57) and (3.61), the static susceptibility is obtained [see Eq. (3.31)]

$$\chi_{ij}^0 \equiv \chi_{ij}[\vec{q}, \omega = 0] = \int_0^\infty \chi_{ij}(\vec{q}, s)\, ds = \Phi_{ij}(\vec{q}, t = 0). \tag{3.63}$$

The spontaneous fluctuations in thermal equilibrium are generalized from Eq. (3.45)

$$C_{ij}(\vec{r}, \omega) = \int_{-\infty}^{\infty} dt\, C_{ij}(\vec{r}, t) \exp(i\omega t)$$

$$\equiv \int_{-\infty}^{\infty} dt\, \langle \Delta A_i \Delta B_j(\vec{r}, t) \rangle \exp(i\omega t), \tag{3.64}$$

where $\Delta A_i = \Delta A_i(\vec{r} = 0, t = 0)$. Using Eq. (3.58) and symmetry properties (3.33) and (3.34) in time, we obtain the fluctuation–dissipation theorem for a nonuniform system

$$\chi_{ij}''(\vec{r}, \omega) = \frac{\beta\omega}{2} C_{ij}(\vec{r}, \omega). \tag{3.65}$$

It relates two physically distinctive quantities of fundamental importance. Spontaneous fluctuations, on the one hand, originate from thermal motion at microscopic level, even in the absence of external field. The irreversible deformation process with the loss of energy, on the other hand, causes the dissipation function of disordered materials.

At this point, it is more illustrative to introduce a causal Green's function defined by

$$G_{ij}(\vec{r}, t) = [C_{ij}(\vec{r}, t)/C_{ij}(\vec{r}, 0)]\theta(t). \tag{3.66}$$

Taking the one-sided Fourier transform in time of this equation yields

$$G_{ij}(\vec{r}, \omega) C_{ij}(\vec{r}, 0) = \int_{0}^{\infty} C_{ij}(\vec{r}, t) \exp(i\omega t)\, dt. \tag{3.67}$$

It is important to point out that $C_{ij}(\vec{r}, \omega)$ in Eq. (3.64) is the two-sided Fourier transform. Combining eqs. (3.64) and (3.67) and noting that $C_{ij}(\vec{r}, t)$ is an even function in time, we get

$$C_{ij}(\vec{r}, \omega) = 2C_{ij}(\vec{r}, 0)\, \mathrm{Re}[G_{ij}(\vec{r}, \omega)]. \tag{3.68}$$

Because of the equivalence among the response, relaxation, and fluctuation functions, $C_{ij}(\vec{r}, 0)$ can be expressed by the static susceptibility in accordance with eqs. (3.25) and (3.63):

$$C_{ij}(\vec{r}, 0) = \chi_{ij}^0/\beta, \quad \text{with } \beta = 1/kT. \tag{3.69}$$

Substituting eqs. (3.68) and (3.69) into Eq. (3.65), we obtain

$$\chi_{ij}''(\vec{r}, \omega) = \omega\chi_{ij}^0\, \mathrm{Re}[G_{ij}(\vec{r}, \omega)]. \tag{3.70}$$

This process is a useful formula that enables us to determine the dissipation function from the Green's function governed by an equation of motion.

3.6   Relaxation Time

As an example to illustrate the formal results presented in the last section, let us consider the diffusion of particles suspended in a fluid. The nature of these particles is arbitrary for the moment. They can be molecules in two-phase fluid or submicron particles in polymer composite. The number of particles does not change with time. Then, their number density $n(\vec{r}, t)$ obeys the continuity equation

$$\frac{\partial n(\vec{r}, t)}{\partial t} + \nabla \cdot \vec{j}(\vec{r}, t) = 0. \tag{3.71}$$

The spatial nonuniform density gives rise to a nonzero current $\vec{j}$. The simplest constitutive equation for the diffusion current is

$$\vec{j}(\vec{r}, t) = -D\nabla n(\vec{r}, t), \tag{3.72}$$

where D is the diffusion constant. Therefore, the causal Green's function satisfies the diffusion equation

$$\frac{\partial G(\vec{r}, t)}{\partial t} = D\nabla^2 G(\vec{r}, t), \tag{3.73}$$

and the boundary conditions

$$G(\vec{r}, t = 0) = \delta(\vec{r}), \qquad G(\vec{r} \to \infty, t) = 0. \tag{3.74}$$

The solution to the above two equations can be obtained via the Fourier–Laplace transformation in time and Fourier transformation in space [see Eq. (3.61)]. The results are

$$G(\vec{q}, \omega) = \frac{1}{-i\omega + Dq^2} \tag{3.75}$$

and

$$G(\vec{q}, t) = \exp(-Dq^2 t), \quad \text{for } t > 0. \tag{3.76}$$

This equation suggests a time scale

$$\tau_L(q) = 1/Dq^2. \tag{3.77}$$

It becomes infinite as $q = 2\pi/\lambda \to 0$, where λ is the wavelength. This unusual behavior of collective fluctuations at large wavelengths is to be appreciated for complex systems. Substituting Eq. (3.75) into Eq. (3.70), we obtain the dissipation function

$$\frac{\chi''(\vec{q}, \omega)}{\omega} = \frac{Dq^2 \chi^0}{\omega^2 + (Dq^2)^2}. \tag{3.78}$$

The right-hand side of this equation tells us to expect a Lorentzian line shape [2] shown in Figure 3.1. Dq^2 gives the width of the Lorentian that is at half maximum

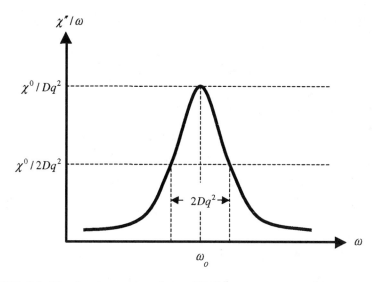

FIGURE 3.1. The imaginary part of the diffusive response over frequency ω [see Eq. (3.78)]. The half width at the half maximum is Dq^2, and a measure of this quantity as a function of q gives the diffusion constant D. The area under the curve gives the static susceptibility.

and has a strong effect on the spectrum of χ''/ω. We have $\omega_o = 0$ in Eq. (3.78). Fourier inversion of Eq. (3.78) gives the familiar expression

$$G(\vec{r}, t) = \frac{1}{(4\pi Dt)^{3/2}} \exp\left(-\frac{r^2}{4\pi Dt}\right), \tag{3.79}$$

which is the Eq. (2.14) for the Gaussian spreading of a random walk.

The next step is to discuss a generalization to the non-Markovian law. Instead of Eq. (3.72), let us look at the constitutive equation with memory

$$\vec{j}(\vec{r}, t) = -\int_{-\infty}^{t} D(t - s)\nabla n(\vec{r}, s)\, ds. \tag{3.80}$$

The function $D(t)$ is called the memory function and is an intrinsic time-dependent property of materials. Let us choose a simplest form [see Eq. (2.75)] with a single structural relaxation time τ. That is

$$D(t) = (D/\tau)\exp(-t/\tau), \tag{3.81}$$

which gives

$$D(\omega) = D/(1 - i\omega\tau). \tag{3.82}$$

This equation introduces frequency dependence into the diffusion coefficient. The relaxation time τ characterizes the microstructure dynamics. By putting all of these dynamics into the continuity equation (3.71), the solution that generalizes Eq. (3.75) is

$$G(\vec{q}, \omega) = \frac{1}{-i\omega + Dq^2/(1 - i\omega\tau)}. \tag{3.83}$$

As we did before, the dissipation function is

$$\frac{\chi''(\vec{q}, \omega)}{\omega} = \frac{Dq^2\chi^0}{\omega^2 + D^2(q^2 - \omega^2\tau/D)^2}. \tag{3.84}$$

When $\omega^2\tau/Dq^2 \ll 1$, Eq. (3.84) reduces to Eq. (3.78). Maxwell originally proposed Eq. (3.79), and Eq. (3.82) is correct both at small and large frequencies.

3.7   The Master Equation

In the study of the irreversible time evolution under the influence of a perturbation, the master equation describes the approach to statistical equilibrium and the transition probabilities of stochastic processes. Let us consider the probability distribution $P_\alpha(t)$ of the system in a state α. The probability distribution changes to $P_\alpha(t + \Delta t)$ for $\Delta t > 0$ in accordance with

$$P_\alpha(t + \Delta t) = \sum_\beta \lambda_{\alpha\beta} P_\beta(t), \tag{3.85}$$

where $\lambda_{\alpha\beta}$ are the transition probabilities [6]. They are normalized as

$$\sum_\beta P_\beta(t) = 1, \qquad \sum_\alpha \lambda_{\alpha\beta} = 1, \tag{3.86}$$

so that we can write

$$P_\alpha(t + \Delta t) - P_\alpha(t) = \sum_\beta \lambda_{\alpha\beta} P_\beta(t) - \left(\sum_\beta \lambda_{\beta\alpha}\right) P_\alpha(t). \tag{3.87}$$

When the transition probabilities per unit time $\Lambda_{\alpha\beta} = \lambda_{\alpha\beta}/\Delta t \geq 0$ are introduced, the familiar form of the master equation is obtained from Eq. (3.87) in the limit of $\Delta t \to 0$:

$$\frac{dP_\alpha(t)}{dt} = \sum_\beta (\Lambda_{\alpha\beta} P_\beta - \Lambda_{\beta\alpha} P_\alpha). \tag{3.88}$$

This expression is often called Pauli's master equation [7,8]. The above equation clearly exhibits the conservation of probability. The first term on the right-hand-side represents the transitions to state α from all other states, and the second sum represents transition from state α to all other states.

One of the most important properties of the master equation is that it has a build-in irreversibility. In nonequilibrium systems, the change in entropy S corresponding to P_α is

$$S(t) = -k \sum_\alpha P_\alpha(t) \log P_\alpha(t). \tag{3.89}$$

Taking the time derivative of this equation, we find

$$\frac{dS(t)}{dt} = -k \sum_\alpha \frac{dP_\alpha(t)}{dt} \log P_\alpha(t) - k \sum_\alpha \frac{dP_\alpha(t)}{dt}. \tag{3.90}$$

Substituting Eq. (3.88) into Eq. (3.90), we obtain

$$\frac{dS(t)}{dt} = -k \sum_{\alpha\beta} [\Lambda_{\alpha\beta} P_\beta(t) - \Lambda_{\beta\alpha} P_\alpha(t)] \log P_\alpha(t). \tag{3.91}$$

By interchanging the names of the dummy summation indexes, Eq. (3.91) can be written as

$$\frac{dS(t)}{dt} = -k \sum_{\alpha\beta} [\Lambda_{\alpha\beta} P_\alpha(t) - \Lambda_{\beta\alpha} P_\beta(t)] \log P_\beta(t). \tag{3.92}$$

Thus, the average is

$$\frac{dS(t)}{dt} = k \sum_{\alpha\beta} [\Lambda_{\alpha\beta} P_\alpha(t) - \Lambda_{\beta\alpha} P_\beta(t)][\log P_\alpha(t) - \log P_\beta(t)] > 0. \tag{3.93}$$

Because $\log P$ is a monotonic increasing function for $P > 0$, every term in the last sum is nonnegative. Whenever the P_α changes, an increase in entropy occurs. The missing information results in entropy increase [8].

The preceding development can be extended in a straightforward manner to the case in which the states form a continuum rather than discrete set. The master equation (3.88) now reads

$$\frac{\partial P(\vec{r}, t)}{\partial t} = \int [\Lambda(\vec{r} \mid \vec{r}') P(\vec{r}', t) - \Lambda(\vec{r}' \mid \vec{r}) P(\vec{r}', t)] \, d\vec{r}', \tag{3.94}$$

where $\Lambda(\vec{r} \mid \vec{r}')$ is the transition probability per unit time jumping from $\vec{r}'$ to $\vec{r}$ and the integration is over the space. We shall see throughout the book that the master equation will be the pivot for understanding and determining the nonequilibrium properties of complex fluids, disordered solids, rough surfaces, and interfaces.

References

1. R. Kubo, M. Toda, and N. Hashitsume, *Statistical Physics II* (Springer-Verlag, Berlin, 1985).

2. D. Forster, *Hydrodynamic Fluctuations, Broken Symmetry, and Correlation Functions* (Benjamin, Reading, MA, 1975).
3. S. W. Lovesey, *Condensed Matter Physics: Dynamic Correlation* (Benjamin, Reading, MA, 1980).
4. L. D. Landau and E. M. Lifshitz, *Statistical Physics* (Pergamon, Oxford, 1969).
5. R. Kubo, Rep. Prog. Phys. 29, Part I, 255 (1966).
6. M. C. Wang and G. E. Uhlenbeck, Rev. Mod. Phys. **17**, 323 (1495).
7. P. M. Mathews, I. I. Shapiro, and D. L. Falkoff, Phys. Rev. **120**, 1 (1960).
8. A. Katz, *Principles of Statistical Mechanics* (Freeman, San Francisco, 1967).

4

Colloidal Dynamics

Colloidal dispersions consist of solid or fluid particles in a solvent. The size of these particles is typically in the mesoscopic range (like Brownian particles) that is much greater than the size of the solvent, thus differentiating these dispersions from molecular solutions. Understanding colloidal dynamics is important in the connection to the rheology of dispersions. Rheology is the science of the flow and deformation of matter, and it exhibits on the one hand by Newtonian viscous fluids and on the other hand by Hookean elastic solids. Most colloidal dispersions exhibit viscoelastic behavior that is intermediate between these two extremes.

The complex flow behavior of colloidal dispersions is sensitive to the mesoscopic structure, which in turn depends on the composition, particle size, shape, orientation, interparticle interactions, and flow field. The flow usually changes from Newtonian for dilute suspensions to non-Newtonian (shear rate dependent) for concentrated dispersions. Experimental data for stable colloids reveal that the effective shear viscosity of dilute and semidilute solutions increases gradually with the volume fraction (ϕ) of particles. However, when the volume fraction ϕ approaches a percolation threshold ϕ_c, a drastic increase in the shear viscosity occurs. This critical phenomenon has been observed for colloidal dispersions, but not for polymer solutions. This phenomenon reflects the fundamental, different mesostructures in colloidal suspensions and in polymeric liquids. In contrast to the long-range interactions in macromolecular systems, the short-range interparticle forces dominate in colloidal dispersions and their magnitude decreases with increasing separation of particles.

Much of the complexity in dispersion rheology will be seen through three types of responses for incompressible shear flows in this chapter. One is time independent, one frequency dependent, and one is time dependent. The viscoelastic

behavior of disperse systems ranging from dilute to semidilute to concentrated
dispersions will be discussed on the basis of statistical dynamics. Contributions
from the hydrodynamic, mesostructure, and interparticle interactions are among
the dominant effects in the analysis of stable hard-sphere dispersions.

We shall start with Stokesian dynamics. The effects of particle shape and flow
orientation on the anisotropic shear viscosities are going to be analyzed. The short-
range mesoscopic structure is particularly important in concentrated suspensions
and will be discussed in terms of a lattice model. This structure helps us to see
the limitation of nanocomposite theories that are based on continuum mechanics. In addition to the pair interactions of colloidal particles that is valid only for
semidilute concentrations, the many-body interactions between the particles and
the equilibrium microstructure in concentrated dispersions have to be included in
the determination of the percolation transition observed in the zero shear viscosity.
The shear thinning is an important non-Newtonian phenomenon and appears in
steady-state shear flow as well as in oscillatory shear flow. We shall discuss the difference between these two flows by distribution function and structural relaxation.
By using the scaling concept of gelation and aggregation expressed by fractals,
a colloid growth model will be introduced to discuss the viscoelastic response of
polymer gels.

4.1 Stokesian Dynamics

Stable colloidal dispersions of nonspherical particle display the effects of particle
shape and flow orientation. To calculate the anisotropic viscosities of a two-phase
system requires a determination of its flow field [1]. Consider a uniform strain rate
tensor, $\dot{e}_{ij}^o$, applied to the system. The total strain rate tensor in the particle or in
the liquid can be written as

$$\dot{e}_{kl}(\vec{r}) = \dot{e}_{kl}^o + \dot{e}_{kl}^C(\vec{r}). \tag{4.1}$$

The dispersed particles add an extra contribution: the local strain rate tensor $\dot{e}_{ij}^C$,
which is a function of coordinates $(\vec{r})$. The corresponding stress tensor in a particle
is

$$\sigma_{ij}^{(1)}(\vec{r}) = C_{ijkl}^{(1)} \big[\dot{e}_{kl}^o + \dot{e}_{kl}^C(\vec{r}) \big], \tag{4.2}$$

where $C_{ijkl}^{(1)}$ is the viscosity tensor of particle. The usual summation convention is
followed for the repeated suffixes over values 1, 2, and 3. A similar expression for
the stress tensor in a liquid medium can also be written simply by replacing $C_{ijkl}^{(1)}$
in Eq. (4.2) with $C_{ijkl}^{(2)}$, the viscosity tensor of liquid. Therefore the bulk stress
tensor is defined as

$$\sigma_{ij}^B(\vec{r}) = \big(C_{ijkl}^{(2)} - C_{ijkl}^{(1)} \big) \big[\dot{e}_{kl}^o + \dot{e}_{kl}^C(\vec{r}) \big], \tag{4.3}$$

which contributes to a body force needed in seeking the solution of the steady-state
shear flow. When $C_{ijkl}^{(1)} = C_{ijkl}^{(2)}$, we expect the trivial solution $\sigma_{ij}^B = \dot{e}_{ij}^C = 0$. For

simplicity, both the particle and liquid phases are assumed to be isotropic. That is [see Eq. (2.32)],

$$C_{ijkl}^{(\alpha)} = \eta_\alpha \left(\delta_{ik}\delta_{jl} + \delta_{il}\delta_{jk} - \tfrac{2}{3}\delta_{ij}\delta_{kl} \right), \qquad \alpha = 1, 2, \tag{4.4}$$

where δ_{ik} is the Kronecker delta and η_1 and η_2 are the shear viscosities of particle and liquid, respectively.

We start with Stokesian dynamics for shear flow in a dilute solution. The total stress tensor and velocity ($\vec{v}$) satisfy the Stokes equation where the inertia and convection terms in the Navier–Stokes equation are neglected (see Appendix 2A). In accordance with, eqs. (4.1) and (4.3), we seek the solution in the form

$$\vec{v} = \vec{v}^o + \vec{v}^C \tag{4.5}$$

and

$$\overset{\leftrightarrow}{\sigma} = \overset{\leftrightarrow}{\sigma}^o + \overset{\leftrightarrow}{\sigma}^C - \overset{\leftrightarrow}{\sigma}^B. \tag{4.6}$$

Using eqs. (4.5) and (4.6) and noting that the applied stress field $2\eta_2\overset{\leftrightarrow}{e}^o$ is uniform, we obtain the equation of motion and continuity, respectively:

$$\eta_2 \nabla^2 \vec{v}^C - \mathbf{grad}\, p = \nabla \cdot \overset{\leftrightarrow}{\sigma}{}^B \tag{4.7}$$

and

$$div\, \vec{v}^C = 0, \tag{4.8}$$

where p is the pressure. The right-hand side of Eq. (4.7) represents a body force. The velocity $\vec{v}^C$ can best be solved by using Green's function $\overset{\leftrightarrow}{G}\,(\vec{r} - \vec{r}')$ governed by

$$\eta_2 \nabla^2 \overset{\leftrightarrow}{G} - \mathbf{grad}\, p = -\overset{\leftrightarrow}{I}\delta(\vec{r} - \vec{r}'), \qquad div\, \overset{\leftrightarrow}{G} = 0, \tag{4.9}$$

where $\overset{\leftrightarrow}{I}$ is the unit tensor and δ is Dirac's delta function. Requiring $\overset{\leftrightarrow}{G}$ approaches zero as $|\vec{r} - \vec{r}'| \to \infty$, we get [1]

$$\overset{\leftrightarrow}{G}\,(\vec{r} - \vec{r}') = \frac{1}{4\pi\eta_2} \left[\frac{\overset{\leftrightarrow}{I}}{|\vec{r} - \vec{r}'|} - \frac{1}{2}\nabla\nabla\,|\vec{r} - \vec{r}'| \right]. \tag{4.10}$$

Equation (4.10) can be written more conveniently as

$$F_{ij}(\vec{r}) = -8\pi\eta_2 G_{ij}(\vec{r}) = -\frac{1}{r}(\delta_{ij} + l_i l_j), \tag{4.11}$$

with

$$r = |\vec{r} - \vec{r}'| \quad \text{and} \quad l_i = (x_i - x_i')/r.$$

Therefore, the solution of eqs. (4.7) and (4.8) is

$$v_i^C(\vec{r}) = -\int_{V_1} \sigma_{jk}^B G_{ij,k}(\vec{r}-\vec{r}')\,d^3\vec{r}' \cong \frac{\sigma_{jk}^B}{8\pi\eta_2}\int_{V_1} F_{ij,k}(\vec{r}-\vec{r}')\,d^3\vec{r}', \quad (4.12)$$

where the bulk stress tensor inside a particle is uniform for dilute suspensions.

The shear rates follow directly from the above equation as

$$\dot{e}_{il}^C = \frac{1}{2}(v_{i,l}^C + v_{l,i}^C) = \frac{\sigma_{jk}^B}{16\pi\eta_2}\int_{V_1}(F_{ij,kl}+F_{lj,ki})\,dV. \quad (4.13)$$

The shear components of bulk stresses and strain rates are related by

$$\sigma_{jk}^B = 2\eta_2 \dot{e}_{jk}^B, \qquad j \neq k. \quad (4.14)$$

The above two equations lead us to introduce a particle-shape tensor $\vec{\vec{S}}$ defined by the equations

$$\dot{e}_{il}^C = S_{iljk}\dot{e}_{jk}^B \quad (4.15)$$

and

$$S_{iljk} = \frac{1}{8\pi}\int_{V_1}(F_{ij,kl}+F_{lj,ki})\,d^3\vec{r}', \quad (4.16)$$

where the reduced Green's function $\vec{F}$ is given by Eq. (4.11) and the integration is performed in the volume of particle, V_1. Eq. (4.16) clearly reveals that the particle-shape tensor depends only on the shape of the particle.

The explicit expressions of the shear components of the particle-shape tensor with $i \neq l$ and $j \neq k$ can be determined by treating the particles as spheroids,

$$\frac{x_1^2}{a^2} + \frac{x_2^2}{a^2} + \frac{x_3^2}{c^2} = 1, \quad (4.17)$$

with the corresponding axes aligned. The medium now becomes anisotropic. The coefficients coupling one shear to another ($S_{1223},\ldots$) are zero. We now have two independent shear components of the particle-shape tensor. Using eqs. (4.11), (4.16), and (4.17), we get

$$S_{1212} = \frac{1}{4}\left(1 - \frac{1-J}{1-\rho^2}\right) \quad (4.18)$$

and

$$S_{1313} = \frac{1+\rho^2}{2}\frac{1-J}{1-\rho^2}, \quad (4.19)$$

where $\rho = c/a$ is the aspect ratio and

$$J = \frac{3\rho}{2(1-\rho^2)^{3/2}}[\cos^{-1}\rho - \rho(1-\rho^2)^{1/2}], \quad \text{for } \rho < 1,$$

$$J = \frac{3\rho}{2(\rho^2-1)^{3/2}}[\rho(\rho^2-1)^{1/2} - \cosh^{-1}\rho], \quad \text{for } \rho > 1.$$

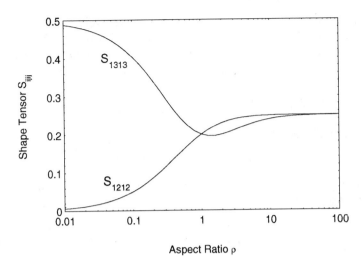

FIGURE 4.1. Shear components of the particle-shape tensor of a spheroid plotted as a function the aspect ratio and shear orientation.

For spherical particles, eqs. (4.18) and (4.19) become

$$\frac{1-J}{1-\rho^2} = S_{1212} = S_{1313} = \frac{1}{5}, \quad \text{for } \rho = 1. \tag{4.20}$$

Figure 4.1 shows the shear components of the particle-shape tensor as a function of the aspect ratio and shear orientation.

4.2   Anisotropic Viscosities   An effective Medium Theory

The effective shear viscosities of a suspension dispersed with nonspherical particles are determined from

$$\eta_{ij} = \frac{\langle \sigma_{ij} \rangle}{2\langle \dot{e}_{ij}^o \rangle}, \quad i \neq j, \tag{4.21}$$

where the angular brackets denote the volume average. No summation over the repeated suffixes is in the above equation and in the rest of this section, unless specified otherwise. When a constant strain rate $\dot{e}_{ij}^o$ is applied to a system at infinity, we have seen in the last section that the perturbed local field is not uniform throughout the fluid composite. The average shear stresses in Eq. (4.21) are given by

$$\langle \sigma_{ij} \rangle = \frac{1}{V} \int_V \sigma_{ij}(\vec{r}) \, dV \quad \text{Liquid} \quad \text{colloid}$$

$$= \frac{1}{V} \left[\int_{V_2} \sigma_{ij} \, dV + \sum \int_{V_1} \sigma_{ij} \, dV \right]$$

$$= \frac{2\eta_2}{V} \int_V \dot{e}_{ij} \, dV + \frac{1}{V} \sum \int_{V_1} (\sigma_{ij} - 2\eta_2 \dot{e}_{ij}) \, dV, \tag{4.22}$$

where V is the total volume and V_1 and V_2 are the volumes of particle and liquid, respectively. The summation is taken over all particles. The stress field inside a particle is related to the bulk stresses by the volume average over a particle:

$$\langle \sigma_{ij}^B \rangle_1 = 2\eta_2 \langle \dot{e}_{ij}^B \rangle_1 = -\frac{1}{V_1} \int_{V_1} (\sigma_{ij} - 2\eta_2 \dot{e}_{ij}) \, dV. \tag{4.23}$$

Thus, Eq. (4.22) becomes

$$\langle \sigma_{ij} \rangle = 2\eta_2 \left(\dot{e}_{ij}^o - \langle \dot{e}_{ij}^B \rangle \phi \right), \tag{4.24}$$

where $\phi = \sum V_1/V$ is the volume fraction. Following eqs. (4.21) and (4.24), the effective viscosities are related to the volume average of bulk stresses by

$$\eta_{ij} = \eta_2 \left[1 - \frac{\langle \dot{e}_{ij}^B \rangle_1}{\dot{e}_{ij}^o} \phi \right], \qquad i \neq j. \tag{4.25}$$

As the volume fraction increases, the interaction between particles becomes important.

We employ a method that is equivalent to effective-medium theory in the study of the particle–particle interaction. When the spatial distribution of the aligned particles is assumed to be random, the dispersed system as a whole has to be macroscopically homogeneous; i.e.,

$$\frac{1}{V} \int_V \dot{e}_{ij}(\vec{r}) \, dV - \dot{e}_{ij}^o = 0, \tag{4.26}$$

where the local $\dot{e}_{ij}$ is given by Eq. (4.1) because of a uniform shear field acting on the system of volume V. Eq. (4.26) can be written more explicitly as

$$\frac{1}{V} \left[\int_{V_2} \dot{e}_{ij}^C(\vec{r}) \, dV + \sum \int_{V_1} \dot{e}_{ij}^C(\vec{r}) \, dV \right] = \phi \langle \dot{e}_{ij}^C \rangle_1 + (1 - \phi) \langle \dot{e}_{ij}^C \rangle_2 = 0, \tag{4.27}$$

where

$$\langle \dot{e}_{ij}^C \rangle_2 = \frac{1}{V_2} \int_{V_2} \dot{e}_{ij}^C(\vec{r}) \, dV. \tag{4.28}$$

The average strain rate $\langle \dot{e}_{ij}^C \rangle_2$ set up in the liquid is a result of the interaction of a particle with its surrounding particles in the case of finite dispersions. When $\phi \to 0$, $\langle \dot{e}_{ij}^C \rangle_2$ should approach zero. Therefore, in the generalization of Eq. (4.15) beyond the dilute limit, it is reasonable to assume that

$$\langle \dot{e}_{il}^C \rangle_1 = S_{iljk} \langle \dot{e}_{jk}^B \rangle_1 + \langle \dot{e}_{il}^C \rangle_2. \tag{4.29}$$

Eliminating $\langle \dot{e}_{ij}^C \rangle_2$ between eqs. (4.27) and (4.29), we obtain

$$\langle \dot{e}_{il}^C \rangle_1 = (1 - \phi) S_{iljk} \langle \dot{e}_{jk}^B \rangle_1. \tag{4.30}$$

In accordance with eqs. (4.1)–(4.3) and (4.14), the average stresses in a dispersed particle have to satisfy the condition

$$C^{(1)}_{ijkl}\left[\dot{e}^{o}_{kl} + \langle \dot{e}^{C}_{kl}\rangle_1\right] = C^{(2)}_{ijkl}\left[\dot{e}^{o}_{kl} + \langle \dot{e}^{C}_{kl}\rangle_1 - \langle \dot{e}^{B}_{kl}\rangle_1\right]. \tag{4.31}$$

The summation over repeated subscripts applies to the above equation. The volume average of the bulk strain rates can be determined from eqs. (4.30) and (4.31), and the effective shear viscosities is then determined from Eq. (4.21) as

$$\frac{\eta_{ij}}{\eta_2} = 1 + \frac{(\eta_1/\eta_2 - 1)\phi}{1 + 2(\eta_1/\eta_2 - 1)(1 - \phi)S_{ijij}}, \qquad i \neq j, \tag{4.32}$$

where the particle-shape tensor is given by eqs. (4.18)–(4.20). By using Eq. (4.32), the dependence of intrinsic viscosities,

$$[\eta_{ij}] = \lim_{\phi \to 0} \frac{\eta_{ij} - \eta_2}{\eta_2 \phi}, \tag{4.33}$$

on the viscosity ratio η_1/η_2 is calculated in Figure 4.2. It clearly reveals that the hard-sphere assumption is valid for $\eta_1/\eta_2 > 20$. For nonspherical particles, however, it may require $\eta_1/\eta_2 > 1000$. In the rest of this chapter, we shall focus our attention on the case of rigid particles, and Eq. (4.32) simplifies to

$$\frac{\eta_{ij}}{\eta_2} = 1 + \frac{1}{2S_{ijij}}\frac{\phi}{1 - \phi}, \qquad \text{for } \eta_1/\eta_2 \gg 1. \tag{4.34}$$

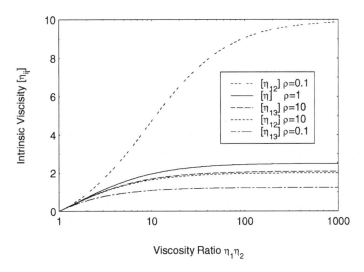

FIGURE 4.2. Dependence of the intrinsic viscosities, defined by Eq. (4.33), on the viscosity ratio between the colloidal particle and liquid. It illustrates the range of validity for the rigid particle assumption in the context of the anisotropic viscosites.

Both eqs. (4.32) and (4.34) have been derived solely on the basis of hydrodynamics. In the case of hard sphere ($\eta_1/\eta_2 \to \infty$ and $\eta_{ij} = \eta$) in dilute suspensions ($\phi \to 0$), eqs. (4.21) and (4.34) yield

$$\eta/\eta_2 = 1 + \tfrac{5}{2}\phi, \tag{4.35}$$

which is the well-known Einstein relation [2].

4.3   Lattice Model

The effective-medium theory developed in the preceding section shows how the effective shear viscosities is influenced by the hydrodynamic interaction between colloidal particles. The short-range mesoscopic interactions within the microstructure will now be discussed (as we shall see later, it is particularly important in concentrated suspensions). Let us look at the free volume or hole theories of the liquid state [3]. Lattice models are basically theories of the solid states and have the advantage of providing good connections between the dense liquid and amorphous solid [4,5; also see Chapter 5].

Consider a lattice consisting of n holes (or free volumes) and n_x polymer molecules of x monomer segments each. The total number of lattice is written in the form

$$N = n + xn_x \quad \text{and} \quad n = \sum_{i=l}^{L} n_i, \tag{4.36}$$

where n_i denotes configurations and L is the total number of hole configurations. When the internal configurations of holes and molecules are neglected, the lattice partition function is

$$Q = \sum_{n_i} W(n_i) \exp\left[-\frac{E(n_i)}{kT}\right], \tag{4.37}$$

where $W(n_i)$ is the number of arrangements of n_i holes in the state energy $E(n_i)$ and n_x molecules on the lattice, k is Boltzman's constant, and T is temperature. The Gibbs free energy of the system is

$$G = -kT \ln Q. \tag{4.38}$$

The equilibrium distribution of the hole number (n_i^*) is determined by minimizing the Gibbs free energy

$$\frac{\partial G}{\partial n_i} = 0, \quad i = 1, 2, \ldots, L, \tag{4.39}$$

which gives

$$\frac{\partial \ln W(n_i)}{\partial n_i} = \frac{1}{kT}\left[\frac{\partial E(n_i)}{\partial n_i}\right] \quad \text{at } n_i = \bar{n}_i. \tag{4.40}$$

The derivative on the right-hand side is called the energy of hole formation in the ith state,

$$\varepsilon_i = \left[\frac{\partial E(n_i)}{\partial n_i}\right]_{n_i = \bar{n}_i}. \tag{4.41}$$

The derivative on the left-hand side is

$$\ln W(n_i) = \Delta S_m / k = -\left[n_x \ln(x n_x / N) + \sum_i n_i \ln(n_i / N)\right], \tag{4.42}$$

where ΔS_m is the entropy of mixing n_i holes and n_x polymer molecules for large L. Combining eqs. (4.40)–(4.42) gives

$$\frac{\bar{n}_i}{N} = c \exp(-\varepsilon_i / kT), \qquad c = \exp(-1 + 1/x). \tag{4.43}$$

The equilibrium hole fraction is

$$\bar{f} = \sum_{i=1}^{L} \bar{n}_i / N = f_r \exp\left[-\frac{\varepsilon}{k}\left(\frac{1}{T} - \frac{1}{T_r}\right)\right], \tag{4.44}$$

where the subscript r refers to an arbitrary reference condition in the liquid state. The mean energy of hole formation, $\varepsilon = \sum_i \varepsilon_i n_i / n$, characterizes the intermolecular interaction.

The shear viscosity of highly viscous fluids is the product of the high-frequency modulus of rigidity (G_∞) and the relaxation time (τ): $\eta = G_\infty \tau$. The instantaneous shear modulus G_∞ can be treated as a constant. The macroscopic relaxation time τ is sometimes called the Maxwellian relaxation time (see Appendix 1A), during which the stresses are damped. This time is also needed for a hole to penetrate barriers in a dense fluid. Thus, we have

$$\eta \propto \tau \propto \exp(\Delta H / kT)), \tag{4.45}$$

and the activation energy (see Chapter 5) is

$$\Delta H = \frac{\varepsilon N}{\beta n} = \frac{\varepsilon}{\beta f}. \tag{4.46}$$

It is related to the local activation energy (ε/f) and $\beta(\leq 1)$, which defines the relaxation-time spectrum (see Chapter 5). Combining eqs. (4.44)–(4.46) yields

$$\frac{\eta}{\eta_r} = \left(\frac{f}{f_r}\right)^{-1/\beta f}. \tag{4.47}$$

When f is chosen to be in the vicinity f_r, Eq. (4.47) can be written approximately as

$$\ln\left(\frac{\eta}{\eta_r}\right) = -\frac{1}{\beta f} \ln\left(\frac{f}{f_r}\right)$$

$$= \frac{1}{\beta f}\left[\left(1 - \frac{f}{f_r}\right) + \frac{1}{2}\left(1 - \frac{f}{f_r}\right)^2 + \cdots\right], \quad \left|1 - \frac{f}{f_r}\right| < 1. \tag{4.48}$$

Therefore,

$$\eta \propto \exp\left(\frac{1}{\beta f}\right), \tag{4.49}$$

which has the form of Doolittle's empirical viscosity equation [4]. This equation relates the viscosity to the free volume. The present lattice model provides us with a useful theoretical basis for a quantitative description of the viscosity of multiphase systems.

4.4   Concentrated Dispersions

We are now in a position to analyze the composition-dependent viscosity of liquid mixtures microscopically. It is assumed that the lattice vibration is unaffected by the composition of the system, and we need only to focus our attention on the configuration changes in the lattice. In accordance with Eq. (4.36), let us consider the number of lattice sites for individual component of binary mixture:

$$N_j(t) = n_j + x_j n_{xj}, \qquad j = 1, 2, \tag{4.50}$$

where n_j and n_{xj} are the number of holes and polymers, respectively, and x_j is the number of monomer segments for the jth polymer component. The lattice is assumed that each molecule occupying a single lattice site with volume v_j. The total volume of the mixture is given by

$$V = vN = \sum_{j=1}^{2} v_j N_j + \Delta V_m, \tag{4.51}$$

where v and $N = n + xn_x$ are, respectively, the lattice volume and total number of lattice sites of the mixture. The excess volume of mixing is ΔV_m. Because the close-packed volumes should remain unchanged, the close-packed volume of mixture is equal to the sum of the pure component close-packed volumes:

$$vxn_x = \sum_{j=1}^{2} v_j x_j n_{xj}. \tag{4.52}$$

Subtracting Eq. (4.52) from Eq. (4.51), we get

$$vn = \sum_{j} v_j n_j + \Delta V_m. \tag{4.53}$$

The volume fraction of the jth component and the excess volume of mixture can be written, respectively, as

$$\phi_j = \frac{v_j N_j}{vN} \quad \text{and} \quad \frac{\Delta V_m}{vN} = A\phi_1\phi_2, \tag{4.54}$$

where A is a nondimensional parameter that measures the strength of the volume interaction between components 1 and 2. From eqs. (4.53) and (4.54), the free volume fraction of the mixture is obtained:

$$f = \frac{vn}{vN} = \sum_j \phi_j f_j + A\phi_1\phi_2. \qquad (4.55)$$

For rigid particles dispersed in liquid, the subscripts 1 and 2 refer to the particle and liquid, respectively. The excess volume is zero and the free volume of particle (f_1) is negligible in comparison with that of liquid (f_2); i.e.,

$$A = 0 \quad \text{and} \quad f_1/f_2 \ll 1. \qquad (4.56)$$

Using eqs. (4.49), (5.55), and (4.56), we obtain

$$\ln\left(\frac{\eta}{\eta_2}\right) \propto \frac{1}{\beta}\left(\frac{1}{f} - \frac{1}{f_2}\right) = \frac{1}{\beta f_2}\frac{\phi}{1-\phi}, \qquad (4.57)$$

where the volume fraction of particles $\phi \equiv \phi_1 = 1 - \phi_2$. Thus,

$$\frac{\eta}{\eta_2} = \exp\left(B\frac{\phi}{1-\phi}\right). \qquad (4.58)$$

Although the focus here is on concentrated dispersions, the liquid lattice model sets no theoretical restriction on the value of ϕ that can be used in the above equation. This helps us to determine the constant B by matching Eq. (4.58) with (4.34) at lower volume concentrations. Expanding the above equation into the series

$$\frac{\overset{\leftrightarrow}{\eta}}{\eta_2} = 1 + \overset{\leftrightarrow}{B}\frac{\phi}{1-\phi} + \cdots, \qquad (4.59)$$

where $\overset{\leftrightarrow}{\eta}$ and $\overset{\leftrightarrow}{B}$ are tensors for nonspherical particles, we obtain

$$\overset{\leftrightarrow}{B} = \frac{1}{2\overset{\leftrightarrow}{S}}. \qquad (4.60)$$

Finally, from eqs. (4.58) and (4.60), the effective shear viscosities of concentrated dispersions are obtained

$$\frac{\eta_{ij}}{\eta_2} = \exp\left[\frac{1}{2S_{ijij}}\frac{\phi}{(1-\phi)}\right], \qquad i \neq j, \qquad (4.61)$$

where the particle-shape tensor S_{ijij} is given in Section 4.1. The effects of the hydrodynamic interactions between particles as well as the mesoscopic interactions within the microstructure have been taken into account in Eq. (4.61). It is a generalization of Eq. (4.34), which is derived on the basis of hydrodynamics alone and is expected to be valid only for lower ϕ. Eq. (4.61) serves as a useful basis in the discussion of the effects of particle shape and orientation on the percolation threshold [6] as ϕ is getting higher.

For simplicity, hard-sphere dispersions will be our focus from now on. The particle-shape tensor reduces to a single constant given by Eq. (4.20), and the effective viscosity $\eta = \eta_{ij}$ is isotropic. Before we move on, we would like to mention that the intermolecular interaction can be estimated from eqs. (4.45), (4.46), (4.57), and (4.58):

$$\varepsilon/f = 2.5\beta kT. \tag{4.62}$$

When the random process for holes is Gaussian, we shall have $\beta = 0.5$ (see Section 4.6). This process results in the interaction potential of $1.25\ kT$ that is very close to the reported value for stable dispersions of hard spheres [7]. In dilute and semidilute dispersions, the expansion of Eq. (4.61) in powers of the volume fraction ϕ is

$$\frac{\eta}{\eta_2} = 1 + 2.5\phi + 5.6\phi^2 + \cdots. \tag{4.63}$$

The coefficient of the ϕ term is 2.5 originally calculated by Einstein for a very dilute suspension. The coefficient 5.6 of the ϕ^2 term consists of 2.5 from Eq. (4.34) as a result of the long-range hydrodynamic interaction of particles and 3.1 from the short-range mesoscopic interaction for the equilibrium microstructure. The coefficient 5.6 in Eq. (4.63) compares well with Batchelor's coefficient of 6.2 [8,9] that decreases to 5.2 if the Brownian motion effect is not taken into account. Eq. (4.61) is derived under the assumption of high shear rate. Figure 4.3 shows

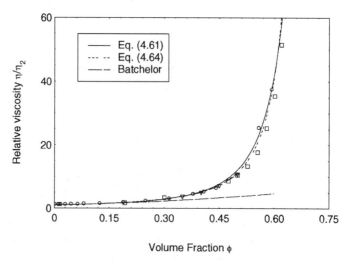

FIGURE 4.3. Present theory [Eq. (4.61) with $\rho = 1$ or Eq. (4.65)] is compared with the experimental data for polystyrene latices in water [10] and silica spheres in cyclohexane [11,12], the Batcheler equation, and an empirical Eq. (4.64). Points are experimental data.

the comparison between Eq. (4.61) and experimental data at the high shear limit of stable colloids ranging from dilute to semidilute to concentrated dispersions. In addition, an empirical relation [13]

$$\frac{\eta}{\eta_2} = \left(1 - \frac{\phi}{0.71}\right)^{-2} \tag{4.64}$$

is also plotted in the figure. The agreement between Eq. (4.64) and Eq. (4.61) with $\rho = 1$ is excellent. Figure 4.3 and Eq. (4.61) reveal that (1) the long-range particle–particle interaction is important in semidilute suspensions, and (2) the short-range mesoscopic interaction with the equilibrium microstructure dominates in concentrate dispersions. We now clearly see the limitation of the continuum mechanics approach in the study of nano-composites containing small particles at higher volume fractions.

4.5 Percolation Transition

Hard-sphere repulsion provides a convenient basis for analyzing the rheological behavior of stable dispersion. The viscosity of dilute suspensions in independent of shear rate ($\dot{\gamma}$) or frequency. As the concentration of colloidal particle increases, the viscosity becomes shear-rate dependent. When contributions from the hydrodynamic and mesoscopic interactions are included in the analysis, the relative shear viscosity of hard-sphere dispersions at a high shear-rate limit has been derived [see Eq. (4.61) with $\rho = 1$] as

$$\eta_r(\infty, \phi) = \frac{\eta_\infty}{\eta_2} = \exp\left(\frac{2.5\phi}{1 - \phi}\right). \tag{4.65}$$

The difference in the limiting viscosities at the low (η_0) and high (η_∞) shear-rate limits is a result of the energy being dissipated through the lost in the kinetic energy (K_E);

$$\Delta\eta \equiv \eta_0 - \eta_\infty = \langle d(K_E)/dt \rangle / \dot{\gamma}^2. \tag{4.66}$$

For particles moving in an inverse-square force field, the classical results $\langle K_E \rangle = -\frac{1}{2}\langle P_E \rangle$ holds, where P_E denotes potential energy. For semidilute suspensions, theories for frequency-dependent shear viscosity have been developed on the basis of pair interactions. When the contribution of all the two-particle interactions in a disperse system is included, Eq. (4.66) becomes [14]

$$\Delta\eta_r = \frac{\Delta\eta}{\eta_2} = \left\langle -\frac{1}{2}\sum_{i>j} r_{ij}\frac{\partial V(r_{ij})}{\partial r_{ij}} \right\rangle, \tag{4.67}$$

where V is the nondimensional pair potential and $r_{ij} = |\vec{r}_i - \vec{r}_j|$ is the distance between the center of two spheres. The summation in the above equation is over all pair interactions. The right-hand side of Eq. (4.67) can be written into an integral

$$-\iint \rho(r_{12}) r_{12} \frac{\partial V(r_{12})}{\partial r_{12}} \, d\vec{r}_1 \, d\vec{r}_2, \tag{4.68}$$

where ρ is the pair distribution function.

The potential energy between two colloidal particles originating from the intermolecular interaction energy (ε), which characterizes the equilibrium microstructure on the basis of the hole theory of liquid lattice mentioned in Section 4.3, is [15,16]

$$U(X) = -\varepsilon/f \equiv -u, \quad \text{for } X < 1$$
$$= 0, \quad\quad\quad\quad \text{for } X \geq 1, \tag{4.69}$$

where $X = r/2a$, f is the hole fraction, and u is called the repulsive interparticle potential. Indeed, the effects of repulsive core dominate the interaction and the longer range forces are washed out for dispersions at finite temperatures. As a result, the negative step function is shown in Eq. (4.69).

The equilibrium radial distribution function is

$$g_o = \exp\left(-\frac{U}{kT}\right). \tag{4.70}$$

For incompressible fluids, we approximate $\rho \sim g_o \psi$, with the distribution function ψ being determined by the Kirkwood–Smoluchowski equation [17] that is a special case of the Fokker–Planck equation:

$$\frac{d}{dX}\left(X^2 g_o \frac{d\psi}{dX}\right) - 6g_o\psi = 4X^3 \frac{dg_o}{dX}. \tag{4.71}$$

Therefore, eqs. (4.67) and (4.68) can be expressed more explicitly as

$$\Delta\eta_r = -\frac{9\phi^2}{5} \int_0^\infty X^3 \frac{dV(X)}{dX} g_o(X)\psi(X) \, dX \equiv A\phi^2, \tag{4.72}$$

where A is the coupling constant. Noting $V = g_o$ in thermodynamic equilibrium, we use eqs. (4.69) and (4.70) and get

$$\frac{dV(X)}{dX} = \frac{dg_o(X)}{dX} = -[\exp(u/kT) - 1]\delta(X - 1). \tag{4.73}$$

The boundary conditions for Eq. (4.71) are that ψ vanishes at infinity and that the pair distribution function is a continuous function. No discontinuity occurs at

$X = 1$. The solution of Eq. (4.71) is

$$\psi(X) = \alpha X^2, \quad \text{for } X < 1$$
$$= \alpha/X^3, \quad \text{for } X \geq 1 \tag{4.74}$$

with $\alpha = 4[\exp(u/kT) - 1][2\exp(u/kT) + 3]$. Substituting eqs. (4.70), (4.73), and (4.74) into Eq. (4.72) yields

$$A = \tfrac{9}{5}[\exp(u/kT) - 1] \int_0^\infty X^3 g_o \psi(x) \delta(X - 1)\, dX = \frac{36[\exp(u/kT) - 1]^2}{5[2\exp(u/kT) + 3]}. \tag{4.75}$$

Eqs. (4.72) and (4.75), derived with the basic elements of pair interaction theories [13,15,17], should be valid for semidilute suspensions.

As the concentration of colloidal particle increases, the many-body interactions between the colloidal particles and the equilibrium microstructure have to be included simultaneously in the determination of the effective zero-shear viscosity, because it is affected by the coupling of the interfering particle to its environment. Consider the probability $P(2, 1)$ of a colloidal particle moving from point 1 to point 2 in a dense dispersion. This probability is the sum of the probabilities of the different ways that the colloidal particle may propagate from 1 to 2. Of course, it interacts with various local lattice sites on its path. For simplicity, we consider only one kind of local site, which has the probability $P(s)$. The total probability is

$$P(2, 1) = P_o(2, 1) + P_o(s, 1)P(s)P_o(2, s)$$
$$+ P_o(s, 1)P(s)P_o(s, s)P(s)P_o(2, s) + \cdots. \tag{4.76}$$

The first term $P_o(2, 1)$ is the probability of free propagation without interference from the local site. In the second term, $P_o(s, 1)$ is the probability of the colloidal particle, which can propagate freely from 1 to site s and $P_o(2, s)$ from s to 2. The third term can be interpreted in the same way. This ideal of the propagation of a particle is from Feynman, and Eq. (4.76) can be represented by the Feynman diagram [18]:

$$(4.77)$$

From the diagram, we can immediately write all various contributions to P, and the interpretation is straightforward. $P_o(2, 1)$ has the Boltzmann-type probability distribution [14]

$$P_o(2, 1) = 1 - \exp(-E_o/kT) \cong E_o/kT + \cdots, \quad \text{for} \quad E_o/kT \ll 1. \tag{4.78}$$

Because $P_o(2, 1)$ is essentially describing the case of a two-particle interaction, E_o can be treated as the pair potential and is related to the nondimensional coupling constant in Eq. (4.72) by

$$E_o \Leftrightarrow A\phi^2. \tag{4.79}$$

Assume the total probability $P(2, 1)$ having the same functional form of Eq. (4.78) and

$$P_o(2, 1) = P_o(s, 1) = P_o(2, s) = P_o(s, s). \tag{4.80}$$

From eqs. (4.72), (4.76), (4.79), and (4.80), we obtain

$$\Delta\eta_r = A\phi^2\{1 + [A\phi^2 P(s)] + [A\phi^2 P(s)]^2 + [A\phi^2 P(s)]^3 + \cdots\}$$
$$= \frac{A\phi^2}{1 - A\phi^2 P(s)}. \tag{4.81}$$

Here, the probability $P(s)$ is produced by the interactions of the colloidal particles with each other and solvent molecules. It may be treated as a fitting parameter of experimental data or be calculated by specifying the structural detail. The many-body interactions during the passage of colloidal particle are analyzed with a set of fixed lattice sites. $P(s)$ is the probability of the particle being "scattered," and $1 - P(s)$ is able to go through. Thus, $P(s)$ is equal to the maximum packing fraction of a unit structure volume filled with spheres.

A percolation threshold is obtained from Eq. (4.81):

$$\phi_c = [A(u)P(s)]^{-1/2} = \frac{\{(5/P)[2\exp(u/kT) + 3]\}^{1/2}}{6[\exp(u/kT) - 1]}, \tag{4.82}$$

where Eq. (4.75) is used. Eqs. (4.81) and (4.82) suggest the disperse systems have a more ordered structure at low shear rate than at high shear rate. Numerically, Eq. (4.82) suggests that the type of ordered arrays in concentrated colloids with either face-centered cubic (fcc: $P(s) = 0.74$) or body-centered cubic (bcc: $P(s) = 0.68$) lattices [19] has very little effect on the dependence of ϕ_c on u/kT. ϕ_c is a decreasing function of the repulsive interparticle potential (u) shown in Figure 4.4, where $P = 0.68$ is used in the calculation for the packing of stable hard spheres. At moderate repulsive potentials, Eq. (4.82) scales simply as $\phi_c \sim (u/kT)^{-1.34}$ for $u/kT \leq 2$. Figure 4.4 reveals a decrease in temperature or an increase in repulsion, which originates with either charged particles at low ionic strength or adsorbing polymer layers, can result in lower ϕ_c. We have determined $u/kT = 1.25$ for a neutrally stable hard-sphere system in Section 4.4, which gives $\phi_c = 0.5733$. This value is very close to the measure critical volume fraction [20].

Combining eqs. (4.65), (4.81), and (4.82) yields the effective zero-shear viscosity

$$\eta_r(0, \phi) = \eta_r(\infty, \phi) + \Delta\eta_r = \exp\left(\frac{2.5\phi}{1 - \phi}\right) + \frac{(\phi/\phi_c)^2}{[1 - \phi/\phi_c)^2]P(s)}, \tag{4.83}$$

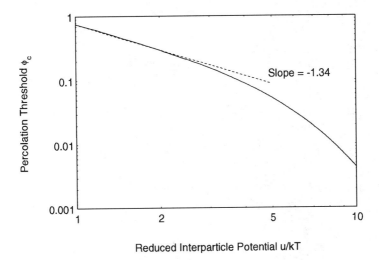

FIGURE 4.4. The relationship between the critical volume fraction ϕ_c and the repulsive inter-particle potential in stable colloidal dispersions.

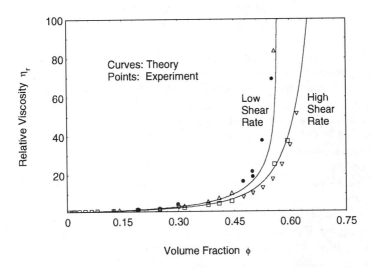

FIGURE 4.5. Calculated high and low limiting shear viscosities, from eqs. (4.65) and (4.83), are compared with experimental data for polystyrene latices in water and silica spheres in cyclohexane [10,11]

which diverges as $\phi \to \phi_c$. Figure 4.5 compares eqs. (4.65) and (4.83) with experimental data of neutral hard-sphere colloids at low and high shear rates ranging from dilute to semidilute to concentrated dispersions. The difference between the low and high curves gets bigger as ϕ increases. Clearly, the shear viscosity is Newtonian for dilute suspensions and becomes non-Newtonian for $\phi > 0.3$.

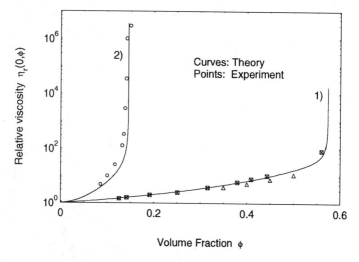

FIGURE 4.6. The zero shear viscosity is plotted as a function of the volume fraction of particles: 1) polystyrene latices in water and silica spheres in cyclohexane [10,11], and 2) polystyrene latices in water at 5×10^{-4} M NaCl [21]. Curves: theory. Points: experiment. The curve and data 1) are the semi-logarithmic plot of the same low shear-rate curve and data shown in Figure 4.5.

Repulsion developing with charged particles at low ionic strengths can result in strong measurable rheological behavior. The potential energy of interactions between colloidal particles is composed of contributions from (1) the hard-sphere repulsion and (2) the electrostatic repulsion related to the interaction of electrical double layers. Several sets of experimental data of polystyrene latices in water at $5 \cdot 10^{-4}$ M NaCl show the same kind of divergence in the zero-shear viscosity of neutral hard-spheres (see Figure 4.5 and curve 1 in Figure 4.6), except the divergence occurs at a much lower percolation threshold $\phi_c = 0.144$ [21]. This rheological similarity is interpreted via Eq. (4.83) by parameters linking to the microstructure (P) and the interparticle potential (u). Using the viscosity data, we find $P(s) = 0.34$ from curve 2 in Figure 4.6, and the ordered array is closed to the description of the diamond lattice. From Eq. (4.82), we obtain $u/kT = 3.74$, consisting of 2.49 from electrostatic repulsion. Colloidal particles have to space out as the repulsion potential increases, which results in lower ϕ_c for charged particles at low ionic strengths.

4.6   Memory Function

Much of the diversity in the viscoelasticity of colloidal dispersions can be seen clearly through an integral equation for simple shear flow. The key to understanding the complex viscosity lies in analyzing the structural relaxation of the microstructure. The relaxation time spectrum and memory function that were mentioned in chapters 2 and 3 are going to play an important role in the calculation of the dynamic viscosity. The linear constitutive equation of the shear stress (σ_{12}) and strain rate ($\dot{\gamma}_{12}$) for a viscoelastic liquid can be written by means of the Boltzmann superposition integral

$$\sigma_{12}(t) = \int_{-\infty}^{t} G(t-s)\dot{\gamma}_{12}(s)\,ds, \tag{4.84}$$

where G is the shear relaxation modulus and t is time. When Eq. (4.84) is subjected to the Fourier transform in time

$$\sigma_{12}[\omega] = \int_{-\infty}^{\infty} \sigma_{12}(t)\exp(-i\omega t)\,dt, \tag{4.85}$$

it can be written in the form

$$\frac{\sigma_{12}[\omega]}{\gamma[\omega]} = G(\omega) = i\omega \int_{0}^{\infty} G(t)\exp(-i\omega t)\,dt, \tag{4.86}$$

where ω is the angular frequency. Please note that the notations for the integral transforms in the above two equations are slightly different from those in chapters 2 and 3. The complex shear viscosity $\eta(\omega)$ is related to the complex modulus by

$$\eta(\omega) = \eta' - i\eta'' = \frac{\sigma_{12}[\omega]}{\dot{\gamma}_{12}[\omega]} = \frac{G(\omega)}{i\omega} = \frac{G' + iG''}{i\omega}. \tag{4.87}$$

Hence,

$$\eta'(\omega) = G''(\omega)/\omega, \qquad \eta''(\omega) = G'(\omega)/\omega. \tag{4.88}$$

As we have learned from Chapter 3, the complex modulus can in general be written as

$$G(\omega) = G_\infty + (G_0 - G_\infty)\Phi(\omega), \tag{4.89}$$

where Φ is the relaxation function, G_0 is the unrelaxed modulus, and G_∞ is the relaxed modulus. In a similar fashion, we can write

$$\eta(\omega) = \eta_\infty + (\eta_0 - \eta_\infty)M(\omega), \tag{4.90}$$

where $\eta_0 = \eta(\omega \to 0)$ and $\eta_\infty = \eta(\omega \to \infty)$ are the limits of viscosity mentioned in the last section and M is the memory function. The above two equations give

$$M(t) = -\tau\frac{d\Phi(t)}{dt}, \tag{4.91}$$

where τ is the relaxation time. It is the ratio of the zero-shear viscosity over the unrelaxed shear modulus: $\tau = \eta_0/G_0$. Eq. (4.91) has exactly the same form of Eq. (3.26), where the response function $\chi = M/\tau$.

The Maxwell model of viscoelasticity gives the simplest relaxation function

$$\Phi_{Maxwell}(t) = \exp(-t/\tau). \tag{4.92}$$

For a real system, Eq. (4.92) has to be generalized

$$\Phi(t) = \sum_i h_i \exp(-t/\tau_i), \tag{4.93}$$

where h_i and τ_i are the distribution and relaxation time, respectively, of the ith relaxation elements. By letting

$$\tau_i/\tau \to s, \qquad h_i \to h(s), \tag{4.94}$$

the summation in Eq. (4.93) can be written as a integral

$$\Phi(t/\tau) = \int_0^\infty h(s) \exp(-st/\tau) \, ds, \tag{4.95}$$

where $h(s)$ is called the relaxation time spectrum. When the distribution is a delta function $h(s) = \delta(s - 1)$, Eq. (4.95) reduces to Eq. (4.92), as expected. From eqs. (4.89)–(4.91) and Eq. (4.95), one gets

$$M(t/\tau) = \int_0^\infty s h(s) \exp(-st/\tau) \, ds. \tag{4.96}$$

On the basis of the free volume or hole theories of liquid state (see Section 4.3), a liquid lattice model has been used to provide a good description of the effective shear viscosity of concentrated colloidal dispersions in Section 4.4. Consider a lattice consisting of n holes defined by Eq. (4.36). The hole configurations change with time in response to shear flow. Let us introduce the local excess of number density $\delta n(\vec{r}, t)$. Then,

$$n(t) - \langle n \rangle = \int \delta n(\vec{r}, t) \, d\vec{r}, \tag{4.97}$$

where $\langle n \rangle$ is the equilibrium value and the integration is taken over a volume element surrounding a hole. The hole density–density correlation function is

$$C(\vec{r}, t) = \frac{\langle \delta n(\vec{r}, t) \delta n(\vec{0}, 0) \rangle}{\langle \delta n^2 \rangle}. \tag{4.98}$$

The local excess of number density relaxes as time goes on by spreading over the entire medium and is governed by the master equation (see Section 3.7)

$$\frac{\partial \delta n(\vec{r}, t)}{\partial t} = \int [\Lambda(\vec{r} \mid \vec{r}') \delta n(\vec{r}', t) - \Lambda(\vec{r}' \mid \vec{r}) \delta n(\vec{r}, t)] \, d\vec{r}', \tag{4.99}$$

where $\Lambda(r \mid r')$ is the transition probability per unit time jumping from $\vec{r}'$ to $\vec{r}$ and the integration is over the whole space. The right-hand side of Eq. (4.99) can be formally expanded into a series [22]. Thus,

$$\frac{\partial \delta n(\vec{r}, t)}{\partial t} = \sum_{m=1}^\infty \frac{1}{m!} (-\vec{\nabla})^m b_m(\vec{r}) \delta n(\vec{r}, t), \tag{4.100}$$

where b_m is the mth moment of the transition rate

$$b_m(\vec{r}) = \int (\vec{r}' - \vec{r})^m \Lambda(\vec{r}' \,|\, \vec{r})\, d\vec{r}, \qquad m \geq 1. \tag{4.101}$$

Because the viscoelasticity of concentrated dispersions is a result of relaxation of the microstructure in quasi-equilibrium, the right-hand side of Eq. (4.100) is truncated after the second term. Eqs. (4.98) and (4.100) give

$$\left[\frac{\partial}{\partial t} - \vec{\nabla} \cdot \frac{b_2}{2} \vec{\nabla}\right] C(\vec{r}, t) = \delta(\vec{r})\delta(t). \tag{4.102}$$

When $b_2/2$ is a (diffusion) constant, the solution of the above equation is

$$C(\vec{r}, t) = \frac{1}{(2\pi b_2 t)^{1/2}} \exp\left(-\frac{x^2}{2b_2 t}\right). \tag{4.103}$$

The Gaussian distribution shown in the above equation has maximum entropy. By considering the structural relaxation in a fixed length scale, the relaxation spectrum can be related to the hole density–density correlation function. We choose $x = 1$ and $b_2 t/2 = s$, replace $C(s)/s$ by $h(s)$ in Eq. (4.103), and obtain

$$h(s) = \frac{1}{2(\pi s^3)^{1/2}} \exp(-1/4s). \tag{4.104}$$

Substituting Eq. (4.104) into eqs. (4.95) and (4.96) gives, respectively, the relaxation function

$$\Phi(t/\tau) = \exp[-(t/\tau)^{1/2}] \tag{4.105}$$

and the memory function

$$M(t/\tau) = \tfrac{1}{2}(t/\tau)^{-1/2} \exp[-(t/\tau)^{1/2}]. \tag{4.106}$$

4.7   Dynamic Viscosities

The dynamic shear viscosity $\eta(\omega, \phi)$ is linked to the memory function by Eq. (4.90):

$$\frac{\eta(\omega, \phi) - \eta_\infty(\phi)}{\eta_0(\phi) - \eta_\infty(\phi)} = M(\omega\tau) = \int_0^\infty M(s) \exp[-is\omega\tau(\phi)]\, ds, \tag{4.107}$$

Substituting Eq. (4.106) into Eq. (4.107), we obtain

$$\begin{aligned}
\frac{\eta(\omega) - \eta_\infty}{\eta_0 - \eta_\infty} &= \frac{\pi^{1/2}}{2(\omega\tau)^{1/2}} \exp\left[\frac{1}{4(\omega\tau)^{1/2}}\right] erfc\left[\frac{1}{2(\omega\tau)^{1/2}}\right] \\
&= \frac{1}{2}\left(\frac{\pi}{2\omega\tau}\right)^{1/2} (1 - i) W\left[\frac{1 + i}{2(2\omega\tau)^{1/2}}\right],
\end{aligned} \tag{4.108}$$

where [23]

$$W\left[\frac{1+i}{2(2\omega\tau)^{1/2}}\right] = \exp\left(\frac{1}{4\omega\tau}\right)\left[1 + \frac{2i}{\pi^{1/2}}\int_0^{(1+i)2(2\omega\tau)^{-1/2}} \exp(-z^2)\,dz\right].$$

(4.109)

The real and imaginary parts of the complex shear viscosity can be written explicitly by expressing Eq. (4.108) by Fresnel integrals

$$C \equiv C[(2\pi\omega\tau)^{-1/2}] = \int_0^{(2\pi\omega\tau)^{1/2}} \cos\left(\frac{\pi}{2}z^2\right)dz$$

(4.110)

and

$$S \equiv S[(2\pi\omega\tau)^{-1/2}] = \int_0^{(2\pi\omega\tau)^{1/2}} \sin\left(\frac{\pi}{2}z^2\right)dz.$$

(4.111)

Therefore,

$$\frac{\eta'(\omega) - \eta_\infty}{\eta_0 - \eta_\infty} = \frac{1}{2}\left(\frac{\pi}{2\omega\tau}\right)^{1/2}\left[(1-2S)\cos\left(\frac{1}{4\omega\tau}\right) - (1-2C)\sin\left(\frac{1}{4\omega\tau}\right)\right]$$

(4.112)

and

$$\frac{\eta''(\omega)}{\eta_0 - \eta_\infty} = \frac{1}{2}\left(\frac{\pi}{2\omega\tau}\right)^{1/2}\left[(1-2C)\cos\left(\frac{1}{4\omega\tau}\right) + (1-2S)\sin\left(\frac{1}{4\omega\tau}\right)\right].$$

(4.113)

The dependence of η' and η'' on $\omega\tau$, calculated from eqs. (4.112) and (4.113), is shown in Figure 4.7, which compares well with the normalized viscosity data [24].

The relaxation time and the zero-shear viscosity can be related by

$$\eta_0(\phi) = G_0 \int_0^\infty \Phi[t/\tau(\phi)]\,dt = 2G_0\tau(\phi),$$

(4.114)

where the unrelaxed shear modulus G_0 is assumed to be much larger than the relaxed modulus, and eqs. (4.91) and (4.106) are again used in Eq. (4.114). The relaxation time is also proportional to the characteristic time of Brownian motion (a^2/D_o). Here, $D_o = kT/6\pi\eta_2 a$ is the Stokes–Einstein diffusion coefficient, and a is the particle radius. Because D_o and G_0 are independent of ϕ, we have

$$\tau(\phi) = \eta_0(\phi)/2G_0 \sim (a^2/D_0)\eta_r(0,\phi), \qquad G_0 \sim kT/a^3,$$

(4.115)

where $\eta_r(0,\phi)$ is given by Eq. (4.83). A comparison between the above equation and the experimental data of hard-sphere colloids is shown in Figure 4.8. We see a sudden increase in $\tau(\phi)$ as ϕ approaches ϕ_c.

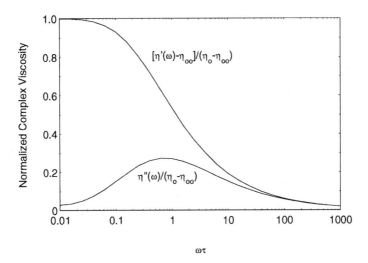

FIGURE 4.7. Calculated master curves of the real and imaginary parts of the complex shear viscosity as a function of $\omega\tau$ for all volume fractions ϕ.

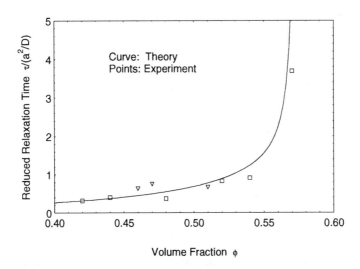

FIGURE 4.8. A comparison of the calculated and measured reduced relaxation time as a function of the volume fraction for silica spheres in cyclohexane [24].

The asymptotic expression of the complex shear viscosity can be obtained from Eq. (4.108), which gives

$$\frac{\eta(\omega, \phi) - \eta_\infty(\phi)}{\eta_0(\phi) - \eta_\infty(\phi)} = 0.627(1 - i)[\omega\tau(\phi)]^{-1/2}, \quad \text{for } \omega\tau \gg 1. \qquad (4.116)$$

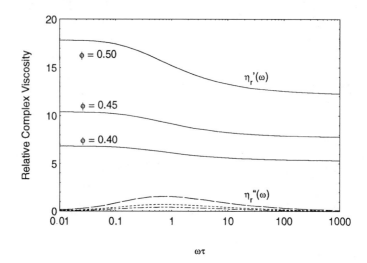

FIGURE 4.9. Calculated real and imaginary parts of the relative complex shear viscosity over the full frequency range with ϕ from 0.40 to 0.50.

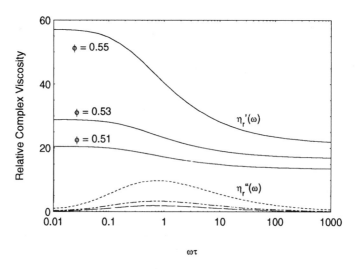

FIGURE 4.10. Calculated real and imaginary parts of the relative complex shear viscosity over the full frequency range with ϕ from 0.51 to 0.55.

The $\omega^{-1/2}$ decay for both the real and imaginary parts of the dynamic viscosity has been supported experimentally [24]. In addition, Eq. (4.116) turns out to be an excellent approximation for $\eta'(\omega)$ starting from the intermediate-frequency range ($\omega\tau \geq 10$) for all ϕ. We are now in the position to calculate the real and imaginary parts of the relative complex shear viscosity over the full ranges of volume fraction and frequency by using eqs. (4.112), (4.113), (4.65), and (4.83). The calculated results are shown in figures 4.9 and 4.10. The lower parts of these two figures

$[\eta_r''(\omega) \equiv \eta''(\omega)/\eta_2]$ correspond to the upper parts $[\eta_r'(\omega) \equiv \eta'(\omega)/\eta_2]$. The highest curve of $\eta_r''(\omega)$ corresponds to the highest volume fraction in each figure. Figures 4.5, 4.9, and 4.10 confirm that the viscoelasticity of stable colloidal dispersions becomes important for $\phi > 0.3$. The viscoelastic effect increases rapidly as $\phi \rightarrow \phi_c$, which is consistent with experimental data [24].

4.8 Mesoscopic Dynamics

Colloidal dispersions have interesting non-Newtonian phenomena, as we have already seen in the last few sections. These phenomena are among the unusual flow behavior that is central to the linear viscoelasticity of oscillatory flow. We shall now give a closer look at how the shear-induced change in microstructure affects the nonlinear flow behavior. The dynamics of disperse systems can be described by the time-dependent distribution function $\psi(\vec{R}_1, \vec{R}_2, \ldots, \vec{R}_N, t)$, where $\vec{R}_i$ is the position of the ith colloidal particle and t is time. The probability (ψ) of finding a particle at a given point and time is governed by the equation of continuity equation [25]

$$\frac{\partial \psi}{\partial t} = -\sum_{i=1}^{N} \frac{\partial}{\partial \vec{R}_i} \cdot \dot{\vec{R}}_i \psi, \qquad (4.117)$$

where $\dot{\vec{R}}_i$ is the velocity of the ith particle. This equation simply states that when a particle leaves one location it has to turn up in another. Eq. (4.117) needs an expression for the velocity $\dot{\vec{R}}_i$, which can be obtained by analyzing the equation of motion of individual particles.

Let us begin with dilute suspensions: The colloidal particle can move freely through the surrounding solvent molecules. Both hydrodynamic and nonhydrodynamic forces are acting on the particle. The equation of motion can be written as

$$m_i \ddot{\vec{R}}_i = -\varsigma_o(\dot{\vec{R}}_i - \vec{v}_i) + \vec{F}_i, \qquad i = 1, 2, \ldots, N, \qquad (4.118)$$

where $\ddot{\vec{R}}_i$ is the acceleration of the ith particle and m_i is the mass. The first term on the right-hand side is the hydrodynamic drag, $\varsigma_o = 6\pi \eta_2 a$ is the friction coefficient, and v_i is the flow velocity that would have existed at the point of location of the ith particle if this particle had been absent. The total nonhydrodynamic force is [8]

$$F_i = -(\partial/\partial \vec{R}_i)(kT \ln \psi + U). \qquad (4.119)$$

The first term is the Brownian force, and the second is the interparticle force with its potential energy U. The acceleration term on the left-hand side of Eq. (4.118) is assumed to be negligible. That is,

$$\ddot{\vec{R}}_i = (1/\varsigma_0)\vec{F}_i + \vec{v}.$$ (4.120)

For concentrated dispersions, the above equation can be generalized to

$$\ddot{\vec{R}}_i = \sum_{j=1}^{N} \ddot{\vec{H}}_{ij}\vec{F}_j + \vec{v}_i,$$ (4.121)

where $\ddot{H}_{ij}$ is the mobility tensor. As mentioned in Eq. (4.118), v_i in the above two equations is caused by the imposed flow field, which is negligible in linear oscillatory shear flow but will be important in the case of nonlinear steady-state shear flow, as we shall see in the next section. Eq. (4.121) accounts for the many-body interactions between particles and solvent molecules in a formal way. Of course, it cannot be solved as it stands, and approximations have to be made to proceed. Combining eqs. (4.117) and (4.121), we obtain

$$\frac{\partial\psi}{\partial t} = \sum_{i=1}^{N}\frac{\partial}{\partial\vec{R}_i}\cdot\left[\sum_{j=1}^{N}\ddot{H}_{ij}\left(kT\frac{\partial\psi}{\partial\vec{R}_j} + \psi\frac{\partial U}{\partial\vec{R}_j}\right) - v_i\psi\right],$$ (4.122)

which is termed the Smoluchowski equation for the disperse system. It may also be called the Fokker–Planck equation.

In practice, Eq. (4.122) will be solved by using an effective one-particle approximation in a self-consistent manner. Physically, it consists of a colloidal particle moving through the surrounding medium whose viscosity is equal to the effective shear viscosity (η) of the dispersion. η has yet to be determined, which allows for a nonlinear response to the interactive system. In the self-consistent model, the mobility tensor is written as

$$\ddot{H}_{ij} = [\ddot{I}/\varsigma(\phi)]\delta_{ij},$$ (4.123)

where $\ddot{I}$ is the unit tensor, and $\varsigma(\phi) = 6\pi a\eta(\phi)$. For dilute solutions in shear [$\varsigma(\phi) = \varsigma_0$], Eq. (4.123) has the exact form used in Rouse's model [26] and in Debye's free draining model. The use of the one-particle self-consistent approximation to solve a many-body problem has been published frequently in various forms for different subjects with great success. This approximation has also been mentioned in the literature as an effective-medium theory. Other examples can be found in Section 4.2 and Chapter 7. The effect of microstructure caused by the many-body interactions mentioned in Eq. (4.121) and the interparticle force in Eq. (4.119) are taken into account collectively in terms of the effective viscosity. Using Eq. (4.123), we obtain

$$\frac{\partial\psi}{\partial t} = \frac{\partial}{\partial\vec{R}}\cdot\left[D\frac{\partial\psi}{\partial\vec{R}} - \vec{v}\psi\right],$$ (4.124)

where $D = kT/\varsigma$ is the effective diffusion coefficient. Both D and ς are functions of η. What we have in Eq. (4.124) is the reduced distribution function of a Brownian particle that is related to that in Eq. (4.117) formally by (see Section 2.4)

$$\psi^{(1)}(\vec{R}_1, t) = \int \psi(\vec{R}_1, \vec{R}_2, \ldots, \vec{R}_N, t) \, d\vec{R}_2 d\vec{R}_3 \cdots d\vec{R}_N. \qquad (4.125)$$

For simplicity, we have dropped the superscript of $\psi^{(1)}$ and subscript of $\vec{R}_1$ on the left-hand side of the above equation as the symbol shown in Eq. (4.124). The shear-induced change in microstructure is described by this reduced particle distribution function in Eq. (4.124) under the influence of a macroscopic velocity field.

4.9 Shear Thinning

Shear thinning is generally believed to result from shear-induced change in microstructure. Experimental data for the effective shear viscosity exhibit two key features: They diverge strongly as the volume fraction (ϕ) of colloidal particles is increased, and they exhibit strong shear-rate ($\dot{\gamma}$) dependence, decreasing with increased shear rate. The shear viscosity is Newtonian for dilute suspensions and becomes non-Newtonian for semidilute suspensions. Shear thinning becomes more pronounced as ϕ approaches a percolation threshold ϕ_c. This unusual nonequilibrium critical phenomenon has been observed for colloidal dispersions, but not for polymer solutions. Although polymer liquids also exhibit non-Newtonian flow behavior, they do not show the critical shear thinning at ϕ_c caused by the long-range interactions in macromolecular systems.

In the development of microscopic theory of the non-Newtonian viscosity, the principal aim is the evaluation of the stress tensor. On the basis of the distribution function mentioned in the last section, we shall analyze the microscopic stress tensor and derive the memory function in steady-state shear flow. By using the method of statistical mechanics, the *configuration average* of the local stress tensor $\sigma_{\alpha\beta}$ created by a given flow field $\vec{v}(\vec{r})$ and Brownian motion in the system is determined by

$$\overline{\sigma_{\alpha\beta}(\vec{r})} = \int \sigma_{\alpha\beta}(\vec{r}, \vec{R}) \psi(\vec{R}) \, d\vec{R}. \qquad (4.126)$$

The spatially dependent stress tensor on the left-hand side of the above equation is equal to the configuration average of particles with stress tensor and the distribution function on the right. Eq. (4.126) has the general form of the statistical average of a physical quantity, which can be a scalar, vector, or tensor [see Eq. (3.8)]. The same form of Eq. (4.126) has also been mentioned in [26]. *Spatial averaging* $\overline{\sigma_{\alpha\beta}(\vec{r})}$ over the total volume V containing N colloidal particles gives the spatial independent stress components:

$$\langle \sigma_{\alpha\beta} \rangle = \frac{1}{V} \int_V \overline{\sigma_{\alpha\beta}(\vec{r})} \, d\vec{r} = \frac{1}{V} \left[\int_{V_2} \overline{\sigma_{\alpha\beta}(\vec{r})} \, d\vec{r} + \sum_{i=1}^{N} \int_{V_i} \overline{\sigma_{\alpha\beta}(\vec{r})} \, d\vec{r} \right], \qquad (4.127)$$

where V_2 are the volume of the solvent. In general, the second-order stress tensor is related to the second-order strain rate tensor by a fourth-order viscosity tensor (see Section 4.2). Because we are only interested in the shear flow, the effective shear viscosity of a disperse system can be obtained from Eq. (4.21). Substituting Eq. (4.127) into Eq. (4.21) and noting that the dispersed system as a whole has to be macroscopically homogeneous, i.e.,

$$\frac{1}{V} \int_V \dot{e}_{\alpha\beta}(\vec{r}) \, d\vec{r} = \dot{e}^o_{\alpha\beta},$$

we obtained the effective shear viscosity of hard-sphere dispersions

$$\eta = \eta_2(1 - \phi) + \frac{\phi \langle \sigma_{\alpha\beta} \rangle_1}{2\dot{e}^o_{\alpha\beta}}, \qquad \alpha \neq \beta, \tag{4.128}$$

where the subscript 1 refers to the colloidal particle and

$$\langle \sigma_{\alpha\beta} \rangle_1 = \frac{1}{V_1} \int_{V_1} \overline{\sigma_{\alpha\beta}(\vec{r})} \, d\vec{r}. \tag{4.129}$$

The hard-sphere fluids is isotropic; i.e., $\eta = \eta_{\alpha\beta}$. Eqs. (4.124), (4.126), and (4.128) together can be regarded as a constitutive equation for a given velocity field. The distribution function is obtained from Eq. (4.124) and the nonlinear viscosity from Eq. (4.128).

Probably the single, most important characteristic of concentrated dispersions is the fact that they have a shear-rate–dependent viscosity. Choose the center of a spherical particle as the origin of spatial coordinates, and consider the steady shear flow

$$\vec{v} = \dot{\gamma} z \vec{j}, \tag{4.130}$$

where $\dot{\gamma} = 2\dot{e}^o_{yz}$ is the shear rate. The inverse of the shear rate is a time scale characterizing the shear thinning and structural relaxation of colloidal dispersions. Substituting Eq. (4.130) into Eq. (4.124), using Eq. (4.126), and noting ψ vanishes at infinity, we find by partial integrations

$$D \frac{d^2 \overline{\sigma_{yz}}}{dy^2} + \dot{\gamma} z \frac{d \overline{\sigma_{yz}}}{dy} = 0. \tag{4.131}$$

The solution of Eq. (4.131) is

$$\overline{\sigma_{yz}(\vec{r})} = A + B \exp\left(-\frac{\dot{\gamma} \varsigma yz}{kT}\right), \tag{4.132}$$

where A and B are constants. Substituting Eq. (4.132) into Eq. (4.129) gives

$$\langle \sigma_{yz} \rangle_1 = A + \frac{B}{V_1} \int_{V_1} \left(1 - \frac{\dot{\gamma} \varsigma yz}{kT} + \cdots\right) d\vec{r}. \tag{4.133}$$

Introducing the polar angles θ and φ,

$$y = r \sin\theta \sin\varphi, \qquad z = r \cos\theta, \tag{4.134}$$

and noting $\dot{\gamma}\varsigma yz$ is the energy induced by the flow field, one obtains

$$\langle yz \rangle_1 = \frac{3}{\pi a^3} \int_0^\pi \int_0^{\pi/2} \int_0^a (r^2 \sin\theta \sin\varphi \cos\theta) r^2 \sin\theta \, dr d\theta d\varphi = \frac{2a^2}{5\pi}. \tag{4.135}$$

The shear-thinning phenomenon has been observed in steady-state shear flow as well as in oscillatory shear flow (see Section 4.7). Therefore, it is important to see the essential difference in the structural relaxation of these two flows. In order to make direct comparison, the shear-rate–dependent effective viscosity derived from eqs. (4.128) and (4.132)–(4.135) is recast in the form of Eq. (4.107) for oscillatory flow:

$$\eta(\dot{\gamma}, \phi) = \eta_\infty(\phi) + [\eta_0(\phi) - \eta_\infty(\phi)] Q[\eta(\dot{\gamma}, \phi)] \tag{4.136}$$

and

$$Q[\eta(\dot{\gamma}, \phi)] = \exp\left[-2.4\phi \frac{a^3 \dot{\gamma}}{kT} \eta(\dot{\gamma}, \phi) \right]. \tag{4.137}$$

Similar to Eq. (4.107), Q is the memory function that goes to unity as $\eta \to \eta_0$, and to zero as $\eta \to \eta_\infty$. No adjusting parameter in Q includes the effects of Brownian motion and shear-rate–dependent structural relaxation. The two constants A and B in Eq. (4.132) are related to the limits of viscosity by

$$\eta_\infty(\phi) = [A\phi + B(1 - \phi)]/\dot{\gamma}$$

and

$$\eta_0(\phi) - \eta_\infty(\phi) = B/\dot{\gamma}.$$

In addition to the shear rate, both η and Q are functions of particle size, concentration, and limits of viscosity. Eqs. (4.136) and (4.137) have to be solved numerically [27]. Figure 4.11 shows the comparison between the calculated results and experimental data for polystyrene spheres of various sizes in different media. The diameter of the hard-sphere varies from 46 to 180 nm. Recently, the Newtonian rheological behavior has been observed for the nanocomposites dispersed with 11-nm particles over wide ranges of concentration and shear rate [28]. This experiment provides an additional confirmation of what we should have expected from Figure 4.11 when $\eta_2 a^3 \dot{\gamma}/kT < 0.001$. We have determined $\phi_c = 0.5733$ in Section 4.5 for a neutrally stable hard-sphere system. Figures 4.12–4.14 are the result of numerical calculation from eqs. (4.136), (4.137), (4.65), and (4.83). In Figure 4.12, the relative shear viscosities at different volume fractions are plotted

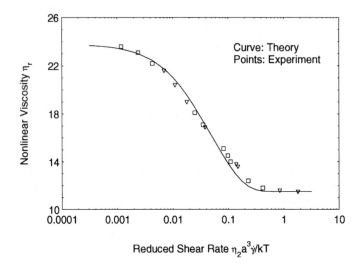

FIGURE 4.11. The calculated relative shear viscosity versus reduced shear rate [from eqs. (4.136) and (4.137)] is compared with the experimental data [10] for mono-dispersions of polystyrene spheres of different sizes dispersed in water, benzyl alcohol, and *meta*-cresol. The volume fraction is $\phi = 0.5$.

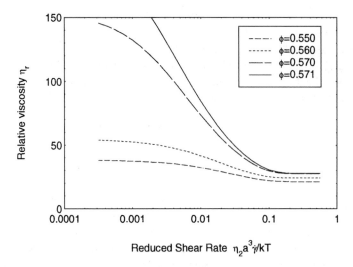

FIGURE 4.12. The nonlinear shear viscosity in steady state shear flow versus the non-dimensional shear rate in the vicinity of percolation threshold.

against the dimensionless shear rate in the vicinity of the percolation threshold. By using Eq. (4.21), the relationships between the shear stress and shear rate are obtained as a function of ϕ in Figure 4.13. From figures 4.12 and 4.13, the dependence of the relative shear viscosity on the reduced shear stress is shown in Figure 4.14, where a dramatic effect of shear thinning in the vicinity of the percolation threshold is seen. This result again is consistent with reported

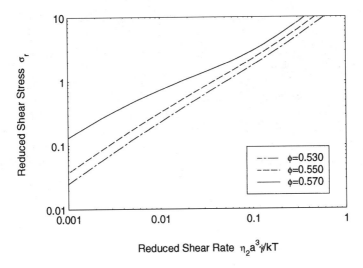

FIGURE 4.13. Nonlinear shear stress and shear rate relationships at different volume fractions of highly concentrated dispersions.

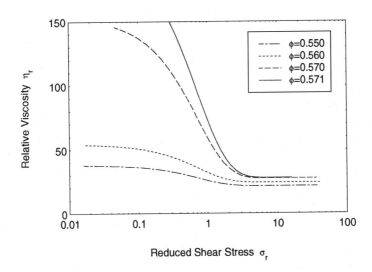

FIGURE 4.14. The nonlinear viscosity in steady state shear flow versus the reduced shear stress in the vicinity of the percolation transition.

measurements [29]. The reduced shear stress is $\sigma_r = a^3 \langle \sigma_{yz} \rangle / kT$. The transition in Figure 4.14 also occurs when σ_r is close to unity. Therefore, the macroscopic shear stress is at least comparable with the instantaneous modulus [see Eq. (4.115)], and a nonlinear flow behavior is expected for the system. It may be worthwhile to remind readers that no adjusting parameter has been introduced in the calculations.

The shear-induced microstructure has a Gaussian distribution for linear oscillatory shear flow [see Eq. (4.103)], but it has a non-Gaussian distribution for nonlinear steady-state shear flow, which results in a fundamental difference between M and Q. The memory functions M in oscillatory shear flow is a function of $\omega\tau$ only. The memory function Q in steady-state shear flow depends not only on the shear rate, particle size, and volume fraction, but also on the effective shear viscosity, which has to be solved simultaneously with eqs. (4.136), (4 137), (4.65), (4.82), and (4.83). Therefore, a time-temperature superposition exists for M (see Figure 4.7), but not for Q, which leads to a familiar classification of viscoelastic behavior: The linear dynamic viscosity in oscillatory shear flow is rheological simple, but the nonlinear viscosity in steady-state shear flow is rheologically complex.

4.10   Colloid Growth Model

The fundamental relationships between the static and dynamic scaling in polymer gels can be derived on the basis of a colloid growth model. Polymer gel is a tenuous random network of particles. Much is already known about critical gels [30]; however, much too little is known about the physics far from the sol–gel transition. Advances in comprehending the relationships between gelation and aggregation in terms of fractals [31,32] should help us to understand the scaling laws in viscoelastic relaxation not only for critical gels, but also for a system with a different degree of crosslinking.

To determine the size dependent viscosity (η) of a system containing branched polymers, it is necessary to understand the dynamics of continuous solid networks surrounded by a continuous liquid phase. Clusters of particles or monomers caused by random aggregation processes form the self-similar structures called fractals (see Appendix 4A). The structures of the self-assembly of a large number of identical elements are between the two extremes: regularity and randomness. With the former, the elements arrange themselves in a periodic fashion, as in the case of crystalline solids. With the latter, one sees nonstructures like those in gases. The fractal dimension (d_f) provides the mass-size scaling:

$$N \sim R^{d_f}, \tag{4.138}$$

where N is the total number of monomers (cluster mass) and R is the radius of gyration (cluster size). Linear polymers can be thought of as linear clusters, and branched polymers can be thought of as branched clusters. Self-avoiding linear polymers have $d_f = 5/3$ for $d = 3$ and $d_f = 4/3$ for $d = 2$. The spatial dimension d affects the range of d_f but not the scaling concepts. We have a very broad range for d_f with either d; however, we shall chose $d = 3$ for most of the numerical discussion. The Flory-excluded volume exponent is equal to $1/d_f = 3/5$ [33]. An ideal phantom network has $d_f = 4$ for highly crosslinked polymers. Thus, branched polymers have $5/3 < d_f < 4$. Here, $d_f = 4$ has been chosen at the value of the critical dimension that defines the upper bound for the density fluctuations in the modeling of gelation [34].

The structural-dependent shear viscosity should be a function of the cluster size and the fractal dimension. In the study of the kinetics of gelation, the density of particles of branched polymers in the region of size R is defined as

$$\rho(R) \sim \frac{N}{V} \sim \frac{R^{d_f}}{R^d} \sim R^{d_f - d}, \tag{4.139}$$

where V is the volume. In order to find the relationship between the shear viscosity and density, a differential equation has to be derived. We analyze the dynamics of branching as a colloid growth process by adding an infinitesimal amount of monomers to a growing cluster in such a way that the radius is increased slightly by ΔR. This idea has been successfully used in the computer simulation of the diffusion-limited gelation [31,32]. Analytically, we introduce

$$\rho(R + \Delta R) = \rho(R) - \Delta\rho \tag{4.140}$$

and

$$\eta(R + \Delta R) = \eta(R) + \Delta\eta \tag{4.141}$$

for an enlarged cluster. The next step is to find the relation of the change in the shear viscosity ($\Delta\eta$) and the change in monomer density ($\Delta\rho$) in eqs. (4.140) and (4.141), respectively, caused by the increase in cluster radius (ΔR) in the limit of $\Delta R \ll R$.

Look at the effective shear viscosity (η) of concentrated hard-sphere suspensions given by Eq. (4.65). It has provided a very good description of experimental data from dilute, semidilute, and concentrated dispersions (see Figure 4.3). In the colloid growth model, the aggregate made of metallic particles has been computer simulated [32,35] to produce a continuous network of particles similar to the structure of polymer gel. In real systems, particles or monomers are not as rigid as metallic solids. It has already been shown in Section 4.2 that the hard-sphere assumption in Eq. (4.65) is a good one as long as the viscosity ratio of particle to its surrounding liquid is greater than 20, which is usually true for polymeric networks.

By following the concept of aggregation kinetics in the study of branching caused by a small variation of polymer concentration, η is replaced by the shear viscosity of the enlarged cluster $\eta(R + \Delta R)$, η_2 by the shear viscosity of the original cluster $\eta(R)$, and ϕ by the density ratio $\Delta\rho/\rho(R)$. Therefore, Eq. (4.65) can be written in the form

$$\eta(R + \Delta R) = \eta(R) \exp\left[\frac{5}{2} \cdot \frac{\Delta\rho}{\rho(R) - \Delta\rho}\right]. \tag{4.142}$$

Combining eqs. (4.140)–(4.142) yields

$$\frac{\eta(R + \Delta R) - \eta(R)}{\eta(r)} \cong -\frac{5}{2} \cdot \frac{\rho(R + \Delta R) - \rho(R)}{\rho(R)}, \quad \text{as } \Delta R \to 0. \tag{4.143}$$

In accordance with the definition of differentiation, the above equation deduces a differential equation for the density dependent shear viscosity:

$$\frac{d\eta}{\eta} = -\frac{5}{2}\frac{d\rho}{\rho}. \tag{4.144}$$

This basic equation is needed in the analysis of branched polymers as continuous solid networks surrounded by a continuous liquid phase. The solution of (4.144) is

$$\eta(R) \sim \rho^{-5/2}(R). \tag{4.145}$$

Substituting Eq. (4.139) into Eq. (4.145) yields the size-dependent viscosity:

$$\eta(R) \sim R^{\frac{5}{2}(d-d_f)} \equiv R^{\kappa}. \tag{4.146}$$

Hence, the viscosity exponent is

$$\kappa = \tfrac{5}{2}(d - d_f), \qquad d > d_f. \tag{4.147}$$

The overlap of the fractal objects is described by d_f in Eq. (4.139). It has been used in the derivation of the above equation, which should be valid for branched polymers with or without overlaps. At the critical point, $d_f = 2.5$ in the diffusion-limited aggregation (DLA; see the next section) case, and $d_f = 2$ in the case of stiff random walks [36].

4.11 Polymer Gels

The shear relaxation modulus (G) is related to the relaxation-time spectrum (H) by [37; also see Section 4.6]

$$G(t) = G_\infty + \int_{-\infty}^{\infty} H(\tau)\exp(-t/\tau)\,d\ln\tau. \tag{4.148}$$

Power-law dependence has been observed over many decades of time for crosslinking systems

$$G(t) \sim t^{-m}, \tag{4.149}$$

where m is the viscoelastic exponent. Alfrey approximated the kernel function $e^{-t/\tau}$ in Eq. (4.148) by a step function going from 0 to 1 at $t = \tau$. This process results in Alfrey's rule for the relaxation-time spectrum:

$$H(\tau) = -\left[\frac{dG(t)}{d\ln t}\right]_{t=\tau} \sim \tau^{-m}. \tag{4.150}$$

The above two equations are valid over decades of intermediate time scales and are within the range of a measurable relaxation-time spectrum. The steady shear viscosity follows:

$$\eta = \int_0^\infty G(t)\,dt \sim \int^t t^{-m}\,dt \sim \tau^{1-m}. \tag{4.151}$$

The relaxation time τ also depends on the self-similar fractal structure of clusters.

The diffusion of a cluster can be described by using a generalized Stokes–Einstein relationship in a medium with a size-dependent viscosity [33]

$$D(R) = \frac{kT}{6\pi\,\eta(R)R} = \frac{D_o(R)}{\eta(R)/\eta_2} \sim \frac{R^{2-d}}{R^\kappa}, \tag{4.152}$$

where η_2 is independent of the cluster size and the diffusion coefficient $D_o = kT/6\pi\,\eta_2 R.$ The size-dependent relaxation time is

$$\tau(R) \sim \frac{R^2}{D(R)} \sim R^{d+\kappa}. \tag{4.153}$$

Substituting Eq. (4.153) into Eq. (4.151), and then comparing it with Eq. (4.146), we get

$$[R^{d+\kappa}]^{1-m} = R^\kappa, \tag{4.154}$$

which yields the viscoelastic exponent:

$$m = \frac{d}{d+\kappa} = \frac{2d}{7d - 5d_f}. \tag{4.155}$$

Because eqs. (4.147) and (4.155) are derived on the basis of a physical picture of the diffusion-limited gelation, these two equations should be valid near the sol–gel transition [30], but they are not valid above the gel point, which we shall see in the following.

On the basis of a nonequilibrium theory of polymeric glasses [see Chapter 5 and Eq. (6.13)], the shear modulus decays in the vicinity of the glass transition as

$$G(t) \sim t^{-[1-\mu(T)]\beta}, \qquad 0 < \mu < 1, \tag{4.156}$$

where the stretched exponent β is independent of temperature and related to the fractal dimension by $\beta = 2/3d_f$. The physical aging exponent μ depends on temperature. The effect of thermal history on glasses is included in $\mu(T)$. We have $\mu < 1$ for $T < T_g$ and $\mu = 0$ for $T > T_g$, where T_g is the glass transition temperature. A detailed analysis and indepth discussions of $\mu(T)$ and β will be given in the next chapter. We are now in the position to introduce a universal constant γ that relates the shear modulus of a viscoelastic gel in melt to that in glass; i.e.,

$$\gamma = (1-\mu)\left[\frac{d\ln G(t)}{d\ln t}\right]_{melt}\bigg/\left[\frac{d\ln G(t)}{d\ln t}\right]_{glass} = \frac{m}{\beta}. \tag{4.157}$$

This universal constant should be valid for networks with different crosslink densities, including the critical gel. We shall determine γ by analyzing the phenomena at the sol–gel transition and then use Eq. (4.157) to describe systems with crosslink densities ranging from linear to crosslinked polymers.

As we mentioned earlier in the derivation of Eq. (4.144), the kinetics of gelation is related directly to that of aggregation, which has also served as the basis in the discussion of the sol–gel transition. DLA is a popular model of forming continuous random networks like branched polymers. The random fractal aggregates have the universal power law in the form of Eq. (138). The sol–gel transition is the critical point at which a system changes from a sol phase (a dispersion of finite molecules) to the gel phase (continuous networks in liquid). The critical fractal dimension has been determined to be $d_{fc} \cong 5d/6 = 2.5$ for $d = 3$ [31] in accordance with the computer-simulated results of metallic particles. This dimension is within the theoretical limits of $5/3 \leq d_{fc} \leq 4$ for linear polymers and highly crosslinked networks, respectively. A remarkable fact is that the rheological properties of a system in the vicinity of d_{fc} are not unlike those near the sol–gel transition.

Substituting $d_{fc} = 2.5$ into Eq. (4.147), we obtain the viscosity exponent of critical gels $\kappa_c = 1.25$. This value is within the experimental limits of 1.4 ± 0.2 for epoxy resins. From Eq. (4.155), the critical viscoelastic exponent is $m_c = 0.706$, which results in $G'(\omega) \sim G''(W) \sim \omega^{0.706}$ at the gel point that is in excellent agreement with the measured complex shear modulus of epoxy resins at 90 °C with $m_c = 0.70 \pm 0.05$. The measured κ_c and m_c [38] had been explained by using percolation models [39] that are in good agreement with the present predictions derived on the basis of a colloid growth model. Using eqs. (4.155)–(4.157), we obtain the universal constant:

$$ \gamma = \frac{3d_{fc}d}{7d - 5d_{fc}} \cong \frac{15}{17}d = 2.647, \quad \text{for } d = 3, \tag{4.158} $$

For systems far away from the gel point, eqs. (4.149), (4.157), and (4.158) are used to interpret the experimental data of crosslinked polystyrenes in melt shown in Figure 4.15. Different values of m represent data at different gel concentrations (v). This figure also reveals that the relaxation time gets longer as v increases, which is in agreement with the discussion given in [26].

The explicit relationships between the crosslink density, the viscoelastic exponent, and the fractal dimension are presented in Table 4.1. By using the universal constant γ given by Eq. (4.158), the fractal dimension can be determined by $d_f = 2\gamma/3m = 1.765/m$ for $d = 3$, where m is obtained from experimental data shown in Figure 4.15. As the gel concentration ($0 \leq v \leq 1$) is increased, Table 4.1 clearly shows that the decrease in the viscoelastic exponent m is caused by the increase in the fractal dimension d_f. At 80% gel, the system has almost approached the state of a highly crosslinked phantom network. Thus, Eq. (4.158) has provided a useful link between the viscoelastic properties of branched polymers in melt and in glass, and its application is not limited to systems in the vicinity of the sol–gel transition. We have used the DLA as a popular example in the discussion of the scaling laws. However, the model is rather general. In the case of stiff random walks ($d_f = 2$), we obtain $m_c = 0.546$: The shear modulus relaxes slower than that of the DLA gels at the sol–gel transition.

TABLE 4.1. *Exponents for Crosslinked Polystyrene (PS)*

Crosslink Density	Viscoelastic Exponent, m	Fractal Dimension, d_f
Linear PS	1.29	1.37
0.40	1.03	1.71
0.80	0.42	4.17

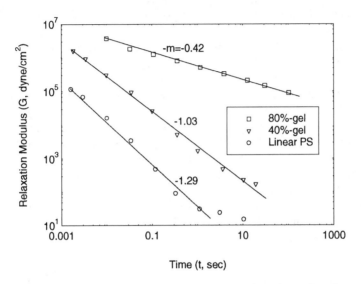

Time (t, sec)

FIGURE 4.15. The effect of crosslink density on the decay of the shear relaxation modulus of crosslinked polystyrenes [40]. Lines: theory. Points: experiment.

As a brief summary of the last two sections, the fundamental relationships between the static and dynamic scaling in polymer gels have been derived on the basis of a colloid-growth model. The cluster size and mass scaling is defined by the static fractal dimension d_f as

$$R \sim N^{1/d_f}, \qquad \text{(static)},$$

and the shear relaxation modulus is obtained:

$$G(t) \sim t^{-2\gamma/3d_f}, \qquad \text{(dynamic)}.$$

The overlap in the fractal objects is also described by d_f. In the DLA case, we have (1) determined the universal constant $\gamma = 2.647$ for $d = 3$ on the basis of the fractal concepts, (2) described the viscoelastic relaxation of crosslinked polymers far from the sol–gel transition in Figure 4.15, and (3) provided the quantitative predictions of the measured viscosity (κ) and viscoelastic (m) exponents at the gel point.

Appendix 4A   Fractals

Since the publication of Mandelbrot's work in 1982 [41], the description of complex fluids and disordered solids has been increasingly couched in fractal language. Although this geometrical concept of fractals has yet to be systematically integrated into physical problems, it is very likely that the fractal descriptions will soon become routine. We shall introduce here the fractal dimension through a simple example. The fractal concept will be used throughout the book to discuss some physical properties of complex fluids, solids, and surfaces in chapters 4, 5, and 8, respectively.

Lines, surfaces, and volumes can be divided into parts of sets of points, lines, and surfaces, respectively, and are commonly attributed to the Euclidean dimensions $d = 1, 2, 3$. Without trying to give a formal definition of self-similarity, let us consider the Sierpinski gasket that is formed in Figure 4A-1, where the first three generations are shown. The subdivisions of the two-dimensional object are the parts identical to each other and are geometrically similar to the original shape. In Figure 4A-1, objects can be either black solids or white holes. After many stages of generation, the black areas will eventually diminished and so will the density.

Start with a triangle with its edge L, area $A(L)$, and mass $M(L)$. When $L \to 2L$, $A(2L) = 2^d A(L)$ and $M(2L) = 2^d M(L)$, where $d = 2$. The density $\rho = M/L$ is not changed. The corresponding expressions for the Sierpinski gasket, however, are $A(2L) = 2^d A(L)$ and $M(2L) = 3M(L) = 2^{d_f} M(L)$. These expressions result in the nonconstant density

$$\rho(2L) = 2^{d_f - d} \rho(L) \tag{4A-1}$$

and the fractal dimension

$$d_f = \ln 3 / \ln 2 = 1.585. \tag{4A-2}$$

In general, for any positive number b, fractal objects scale as

$$M(bL) = b^{d_f} M(L). \tag{4A-3}$$

The solution of this equation is obtained by setting $b = 1/L$:

$$M(L) \sim L^{d_f}, \tag{4A-4}$$

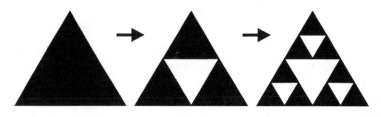

FIGURE 4A-1 Sierpinski gasket is approached after three stages of iteration.

which is a useful relation for the mass-length scaling. Hence,

$$\rho(L) \sim L^{d_f - d}.\qquad(4A\text{-}5)$$

The density correlation function $\langle p(\vec{r} + \vec{L})\rho(\vec{r})\rangle$ is another useful relation in which $\vec{r}$ is the position vector. The density at distance L from an origin point is the ratio of mass to volume in a spherical shell of size proportional to L^{d-1}. From Eq. (4A-5), it also follows that

$$\langle \rho(L)\rho(0)\rangle \sim L^{d_f - d}.\qquad(4A\text{-}6)$$

References

1. L. D. Landau and E. M. Lifshitz, *Fluid Mechanics* (Addison-Wesley, Reading, MA, 1959).
2. A. Einstein, Ann. Phys. **17**, 549 (1905); **19**, 289 (1906); **34**, 591 (1911).
3. T. L. Hill, *Statistical Mechanics* (Dover, New York, 1987).
4. A. K. Doolittle, J. Appl. Phys. **22**, 1471 (1951).
5. T. S. Chow, Adv. Polym. Sci. **103**, 149 (1992).
6. T. S. Chow, Phys. Rev. E **48**, 1977 (1993).
7. R. Buscall, J. Chem. Soc., Faraday Trans., **87**, 1365 (1991).
8. G. B. Batcherler, J. Fluid Mech. **83**, 97 (1977).
9. G. B. Batcherler and J. T. Green, J. Fluid Mech. **56**, 401 (1972).
10. I. M. Krieger, Adv. Colloid Interface Sci. **3**, 111 (1972).
11. C. G. de Kruif, E. M. F. van Iersel, A. Vrij, and W. B. Russel, J. Chem. Phys. **83**, 4717 (1985).
12. J. C. Van der Werff and C. G. de Kruif, J. Rheol. **33**, 421 (1989).
13. W. B. Russel, D. A. Saville, and W. R. Schowalter, *Colloidal Dispersions* (Cambridge University Press, New York, 1989).
14. R. P. Feynman, *Statistical Mechanics* (Benjamin, Reading, MA, 1972).
15. B. Cichocki and B. U. Felderhof, Phys. Rev. A **46**, 7723 (1992).
16. T. S. Chow, Phys. Rev. E **50**, 1274 (1994).
17. J. D. Green, P. F. Buff, and M. S. Green, J. Chem. Phys. **17**, 988 (1949).
18. G. Rickayzen, *Green's Functions and Condensed Matter* (Academic, New York, 1980).
19. C. Kittel, *Introduction to Solid State Physics*, 6th ed. (Wiley, New York, 1986).
20. P. D'Haene, G. G. Fuller, and J. Mewis, *Theoretical and Applied Rheology*, Vol. 2, P. Moldenaers and R. Keunings, editors (Elsevier, Amsterdam, 1992).
21. R. Buscall, J. W. Goodwin, M. W. Hawkins, and R. H. Ottewill, J. Chem. Soc., Faraday Trans. I, **78**, 2872 (1982).
22. J. E. Moyal, Royal Stat. Soc. B **11**, 150 (1949).
23. M. Abramowitz and I. A. Stegun (Eds.), *Handbook of Mathematical Functions* (U. S. GPO, Washington, DC, 1964).
24. J. C. Van der Werff, C. G. de Kruif, C. Blom, and J. Mellema, Phys. Rev. A **39**, 795 (1989).
25. R. B. Bird, R. C. Armstrong, and O. Hassager, *Dynamics of Polymeric Liquids*, Vol. 2, 2nd ed. (Wiley, New York, 1987).
26. M. Doi and S. F. Edwards, *The Theory of Polymer Dynamics* (Clarendon, Oxford 1986).

27. T. S. Chow, J. Phys. Condens. Matter **8**, 8145 (1996).
28. L. Vekas, private comunication (1998).
29. R. Buscall, *Theoretical and Applied Rheology*, Vol. 2, P. Moldenaers and R. Keunings, editors (Elsevier, Amsterdam, 1992).
30. H. H. Winter, Mater. Res. Soc. Bull. **16**(8), 44 (1991).
31. H. E. Stanley, F. Family, and H. Gould, J. Poly. Sci., Polym. Symp. **73**, 19 (1985).
32. T. A. Witten, Jr., and L. M. Sander, Phys. Rev. **47**, 1400 (1981).
33. P. G. de Gennes, *Scaling Concepts in Polymer Physics* (Cornell, Ithaca, 1979).
34. H. E. Stanley, "Critical phenomena," in *Encyclopedia of Polymer Science*, Vol. 4, 2nd ed. (Wiley, New York, 1986).
35. P. Meakin, Phys. Rev. B **27**, 604 (1983).
36. M. E. Cates and S. F. Edwards, Proc. R. Soc. A **395**, 226 (1984).
37. J. D. Ferry, *Viscoelastic Properties of Polymers*, 3rd ed. (Wiley, New York, 1980).
38. J. E. Martin, D. Adolf, and J. P. Wilcoxon, Phys. Rev. Lett, **61**, 2620 (1988).
39. D. Stauffer, *Introduction to Percolation Theory* (Taylor and Francis, London, 1985).
40. T. S. Chow, Macromol. Theory Simul. **7**, 257 (1998).
41. B. B. Mandelbrot, *The Fractal Geometry of Nature* (Freeman, New York, 1982).

5

Glassy–State Relaxation

Amorphous solids are not in thermodynamic equilibrium. The experiment of volume relaxation below the glass transition temperature has revealed that the glassy states are indeed changing slowly with time and temperature [1,2]. The glassy-state relaxation is a result of the local configuration rearrangement of molecular segments, and the dynamic of holes (free volumes) provides a quantitative description of segmental mobility. On the basis of the statistical dynamics of hole motion, a unified physical picture emerges that enables us to discuss the structural relaxation, physical aging, and glassy-state deformation (see Chapter 6). The physics of glassy polymers is still evolving, and the functional relationships between relaxation and deformation have not been firmly established. Significant progress, however, has already been made that can be used in solving many important problems in this area.

We shall start with the concept of free volume and the equilibrium liquid state. It is then followed by a statistical dynamic theory of glasses in which the nonequilibrium glassy state, relaxation time, and relaxation-time spectrum will be discussed. These issues provide us with a solid foundation for a quantitative description of the volume relaxation, recovery, and the pressure–volume–temperature (PVT) behavior of amorphous polymers in both the equilibrium liquid and glassy states. We shall see that the thermal history behavior and the memory effects will emerge as the key features in the understanding of the physical properties of glass-forming polymers.

5.1 Equilibrium State

In liquids, the concept of free volume (hole) has proved to be useful in discussing the viscosity (see sections 4.3 and 4.4). The interpretation of hole, however, is far from unanimous, especially in the glassy state. Let us consider a lattice consisting of n holes and n_x polymer molecules of x monomer segments each. The total number of lattice sites is written in the form

$$N(t) = n(t) + xn_x, \tag{5.1}$$

where $n \ll N$. It is important to mention that n consists of both equilibrium and nonequilibrium contributions in the glassy state. For temperature above the glass transition temperature (T_g), the nonequilibrium contributions to n go to zero. The change in hole population with time (t) and temperature (T) below T_g determines the relaxation processes in the glassy state. Minimizing the excess Gibbs free energy caused by the hole introduction, with respect to the hole number, the temperature dependence of the equilibrium hole fraction is given by (see Section 4.3)

$$\bar{f}(T, p) = \frac{\bar{n}}{N} = f_r \exp\left[-\frac{\varepsilon + pv}{k}\left(\frac{1}{T} - \frac{1}{T_r}\right)\right], \tag{5.2}$$

where ε is the mean energy of hole formation, p is the pressure, v is the lattice volume [see Eq. (6.48)], and the subscript r refers the condition at $T = T_r$. It is a fixed temperature near T_g, which on the other hand is sensitive to the measuring time scales, i.e., cooling rate in the glass-forming process, and strain rate in deformation. The "bar" in Eq. (5.2) refers to the equilibrium of a fully relaxed state with the nonequilibrium hole fraction,

$$\delta(T, p, t) = \frac{n(t) - \bar{n}}{N} = f(T, p, t) - \bar{f}(T, p), \tag{5.3}$$

equal to zero. Eq. (5.2) reveals that holes are created by raising the temperature and are eliminated by lowing it. The microstructure of polymeric materials, such as the chain conformation (see Section 6.1), can be included in the analysis of Eq. (5.1) but has been shown to have little effect on Eq. (5.2).

When each hole occupies only a single lattice cell with volume v, the total macroscopic volume is

$$V(T, p, t) = v(T, p)N(T, p, t). \tag{5.4}$$

The occupied volume vxn_x is actually the molecular volume of polymers, and vn is the hole or free volume. In the state of equilibrium $(\delta = 0)$, the total thermal expansion of the system is

$$\bar{\alpha} = \frac{1}{\bar{V}}\left(\frac{\partial \bar{V}}{\partial T}\right)_p = \frac{1}{v}\left(\frac{\partial \bar{v}}{\partial T}\right)_p + \left(\frac{\partial \bar{f}}{\partial T}\right)_p. \tag{5.5}$$

The first term on the right-hand side is the thermal expansion coefficient (α_v) of the occupied volume and has nothing to do with the glass transition phenomenon. It is independent of temperature. The second term is the excess thermal expansion coefficient of the free volume $(\Delta\alpha)$ of liquid with respect to that of glass near T_g.

Using Eq. (5.2), we get

$$\Delta\alpha = (\bar{\alpha} - \alpha_v)_{T=T_r} = \left(\frac{\partial \bar{f}}{\partial T}\right)_{p,T=T_r} = \frac{(\varepsilon + pv)f_r}{kT_r^2}. \tag{5.6}$$

In the same manner, the excess isothermal compressibility is

$$\Delta\kappa = -\left(\frac{\partial \bar{f}}{\partial p}\right)_{T,T=T_r} = \frac{vf_r}{kT_r}. \tag{5.7}$$

Eqs. (5.6) and (5.7) lead to

$$\frac{\Delta\alpha}{\varepsilon + pv} = \frac{\Delta\kappa}{T_r v}. \tag{5.8}$$

More explicitly, the total equilibrium volume of the system without external pressure ($p=0$) is given by

$$\bar{V} = V_r \frac{1 + \alpha_v(T - T_r)}{1 - \bar{f}}, \tag{5.9}$$

where $V_r = x n_x v_r$ and v_r is the lattice volume at $T = T_r$ close to T_g. When the experimental $V - T$ data of poly(vinyl acetate) (PVAc) above $T_r = 308$ K at atmospheric pressure are used, eqs. (5.2) and (5.9) give $V_r = 0.817$ cm^3/gm, $\alpha_v = 2.1 \times 10^{-4}$/K, $\varepsilon = 2.51$ kcal/mol, and $f_r = 0.0336$ [3]. The slope of $V - T$ data equals $\alpha_v + \Delta\alpha$ and determines $\Delta\alpha = 4.7 \times 10^{-4}$/K, which is close to the "universal" average value of 4.8×10^{-4}/K for the excess thermal expansion of free volume [1]. These parameters will be used later in the calculations of the physical properties of PVAc below T_r. A comparison between the theory and experiment is shown in Figure 5.1.

5.2   Free-Volume Distribution

Positron annihilation life (PAL) spectroscopy has emerged as the unique probe for free volume at molecular and atomic scales (a few Ångstrom in size). The average hole size and volume distributions are found to be dramatically affected by temperature and pressure. Building on the lattice model, we shall include the density fluctuations of holes in the analysis because it will lead to the prediction of many important features of the free volume distributions in amorphous polymers observed in the PAL experiments.

Away from the equilibrium average but still in uilibrium states, a Taylor expansion about $\bar{n}$ of the enthalpy of a polymeric system is

$$H(n) = H(\bar{n}) + \left(\frac{\partial H}{\partial n}\right)_{n=\bar{n}} (n-\bar{n}) + \cdots = H(\bar{n}) + (\varepsilon + pv)(n-\bar{n}) + \cdots. \tag{5.10}$$

The second term in the above equation is the fluctuation ter1   The effect of $(n - \bar{n})$ is ignored in the last section, but it is important in the study   the hole distribution

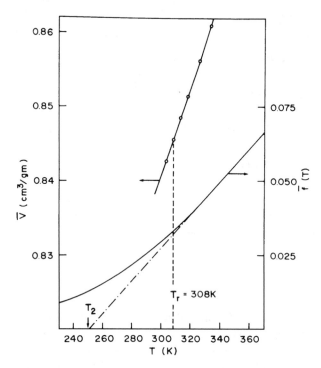

FIGURE 5.1. Plots of the equilibrium volume and free volume fraction versus temperature for PVAc, where circles are experimental data [2].

function, $g(n)$, in accordance with the partition function given by Eq. (4.37), where the energy E has been replaced by the enthalpy H:

$$Q = \sum_n W(n) \exp\left[-\frac{H(n)}{kT}\right] \equiv \sum_n g(n). \qquad (5.11)$$

Thus,

$$\frac{g(n)}{g(\bar{n})} = \frac{W(n)}{W(\bar{n})} \frac{\exp\{-[H(\bar{n}) + (\varepsilon + pv)(n - \bar{n})]/kT\}}{\exp[-H(\bar{n})/kT]}$$

$$= \frac{W(n)}{W(\bar{n})} \exp\left[-\frac{(\varepsilon + pv)(n - \bar{n})}{kT}\right]. \qquad (5.12)$$

Substituting Eq. (5.2) into Eq. (5.12), we get

$$\ln\left[\frac{g(n)}{g(\bar{n})}\right] = N \ln\left(\frac{N}{\bar{N}}\right) - n \ln\left(\frac{n}{\bar{n}}\right). \qquad (5.13)$$

Assume that the density fluctuations in the system are small; i.e.,

$$|y| = \left|\frac{n - \bar{n}}{\bar{n}}\right| \ll 1. \qquad (5.14)$$

Eq. (5.13) can be written approximately as

$$\ln\left[\frac{g(n)}{g(\bar{n})}\right] = \bar{n}[y - (1 + y)\ln(1 + y)] = \bar{n}\left(-\frac{y^2}{2} + \frac{y^3}{6} + \cdots\right). \qquad (5.15)$$

The leading term in Eq. (5.15) results in the Gaussian distribution

$$g(n) = \frac{1}{(2\pi\bar{n})^{1/2}} \exp\left[-\frac{(n - \bar{n})^2}{2\bar{n}}\right]. \qquad (5.16)$$

In practice, the assumption given by Eq. (5.14) is too restrictive for most polymer applications. We must return to Eq. (5.13), from which the non-Gaussian distribution function is derived:

$$g(n) = g_o\left(\frac{n}{\bar{n}}\right)^{-n} \exp(n - \bar{n}), \qquad (5.17)$$

where

$$g_o = \left[\int_0^\infty (n/\bar{n})^{-n} \exp(n - \bar{n}) \, dn\right]^{-1}. \qquad (5.18)$$

At this point, it is easy to show that the equilibrium average is

$$\bar{n} = \int_0^\infty ng(n) \, dn. \qquad (5.19)$$

Eq. (5.17) is derived from the energetic consideration in which the enthalpy of hole formation defines the creation and annihilation of holes as a function of temperature and pressure.

In terms of free volume $V_f = vn$, Eq. (5.17) can be written in the form

$$g(V_f) = g_o\left(\frac{V_f}{\bar{V}_f}\right)^{-\bar{n}V_f/\bar{V}_f} \exp\left[\bar{n}\left(\frac{V_f}{\bar{V}_f} - 1\right)\right], \qquad (5.20)$$

where $\bar{V}_f = \bar{n}v$ is the average equilibrium hole volume that occurred at the most probable distribution Max [g]. The lattice volume v is not a constant, as we mentioned in the last section. It may be treated, however, as a constant in the first approximation. By using $\varepsilon = 2.29$ kcal/mol and $v = 11.5 \text{Å}^3$ for an epoxy polymer with $T_g = 62\,^\circ$C [4], the hole volume distributions are calculated from Eq. (5.20) as a function of temperature and pressure in figures 5.2 and 5.3, respectively. An increase in temperature broadens the distribution and shifts its lower peak to higher free volume. The most probable hole size increases from 68 Å^3 at 100 $^\circ$C to 98 Å^3 at 150 $^\circ$C. Increase in pressure has an opposite effect. The calculated results compare well with the PAL experiments [5]. The present discussion is limited to temperatures above T_g. In the case of $T < T_g$, the effect of physical aging is expected to play a significant role in nonequilibrium glasses.

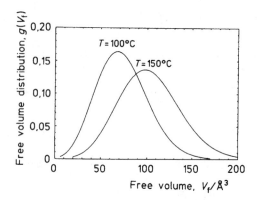

FIGURE 5.2. Effect of temperature on the hole distribution function of an epoxy resin.

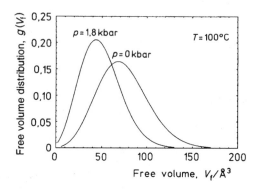

FIGURE 5.3. Effect of pressure on the hole distribution function of an epoxy resin.

5.3   Fractal Dynamic Theory of Glasses

For high molecular weight polymers, the glassy-state relaxation is a result of local configuration rearrangements of molecular segments, which are described by the hole motion and bond rotation. They are expressed by the hole and conformational energies. It has been reported that the conformational activation energy controlling the hindered rotational relaxation for bonds of main chain in the macromolecule is between 1 and 2 orders magnitude lower than the local activation energy for holes [see Eq. (5.37)]. For PVAc, the former is 1.94 kcal/mol and the latter is 74.7 kcal/mol, which are the barriers for local kinetics [6]. As a result, the conformer relaxes much faster than the hole. Because all physical properties of glasses vary slowly in time (t), the dominant contribution to the structural relaxation and physical aging in glasses is from the hole. When the polymer is cooled from liquid to glass where the sample is annealed, the hole configuration space in the quenched and annealed glass is divided into regions separated by barriers. To include the spatial vector ($\vec{r}$) into the analysis, $n(t) - \bar{n}$ is related to the local excess of hole number density $\delta n(\vec{r}, t)$ by integrating over the volume surrounding the individual hole

$$n(t) - \bar{n} = \int \delta n(\vec{r}, t) \, d\vec{r}. \tag{5.21}$$

Because of the slow-varying properties, polymeric glass is considered in a state of quasi-equilibrium. The number of holes does not change much during isothermal annealing. The excess local density of a quenched glass relaxes by spreading slowly over the entire region in accordance with the Master equation (see Section 3.7)

$$\frac{\partial \delta n(\vec{r}, t)}{\partial t} = \int [\Lambda(\vec{r} \mid \vec{r}') \delta n(\vec{r}', t) - \Lambda(\vec{r}' \mid \vec{r}) \delta n(\vec{r}, t)] \, d\vec{r}', \tag{5.22}$$

where $\Lambda(\vec{r} \mid \vec{r}')$ is the transition probability per unit time jumping from $\vec{r}'$ to $\vec{r}$ and the integration is over the space. The right-hand side of Eq. (5.22) can be formally expanded into series

$$\frac{\partial \delta n(\vec{r}, t)}{\partial t} = \sum_{m=1}^{\infty} \frac{1}{m!} (-\vec{\nabla})^m b_m(\vec{r}) \delta n(\vec{r}, t), \tag{5.23}$$

where

$$b_m(\vec{r}) = \int (\vec{r}' - \vec{r})^m \Lambda(\vec{r}' \mid \vec{r}) \, d\vec{r}, \qquad m \geq 1. \tag{5.24}$$

Because all properties of glass vary slowly in space and time, the right-hand side of Eq. (5.23) is truncated after the second term, which gives

$$\sum_{m=1}^{\infty} \frac{1}{m!} (-\vec{\nabla})^m b_m(\vec{r}) \delta n(\vec{r}, t) \approx \vec{\nabla} \cdot D \vec{\nabla} \delta n(\vec{r}, t) + \cdots,$$

where $D = b_2/2$ is the local diffusion coefficient. We have assumed here that no external field is applied to the system. The motion of holes in response to molecular fluctuation can then be treated as an anomalous diffusion by

$$\left(\frac{\partial}{\partial t} - \vec{\nabla} \cdot D \vec{\nabla} \right) \delta n(\vec{r}, t) = 0. \tag{5.25}$$

The initial condition is that δn is nonzero only at $\vec{r} = 0$. When D is a constant, the solution of Eq. (5.25) displays the well-known Gaussian spreading. We have a spatial-dependent diffusion coefficient, however. Let us introduce the Fourier transform in space,

$$\delta n(\vec{Q}, t) = \int \delta n(\vec{r}, t) \exp(-i \vec{Q} \cdot \vec{r}) \, d\vec{r}, \tag{5.26}$$

where $\vec{Q}$ is the wave vector of the fluctuation. Thus, Eq. (5.25) is generalized to the form [7]

$$\left(\frac{\partial}{\partial t} - DQ^{2+\nu}\right)\delta n(\vec{Q}, t) = 0,$$ (5.27)

where ν produces a fractal dimension d_h, which defines a self-similar scaling between wavenumbers by the following transformation function:

$$Q \sim q_\nu^{d_h}, \quad \text{with} \quad d_h = \frac{2}{2+\nu}.$$ (5.28)

Fractal is by definition a set made of parts similar to the whole [8, Appendix 4A]. Self-similarity is the basic notion in fractal structure and a common feature for all scaling analysis. Therefore, the fractal dimension d_h is introduced in accordance with the spatial scaling [Eq. (5.28)]. The $Q^{2+\nu}$ dependence in Eq. (5.27) is an ansatz because $\vec{r}$ dependence of D in Eq. (5.25) is not known. The consequences of this ansatz are broad and far reaching. Using eqs. (5.27) and (5.28), we shall be able to derive the kinetic equations [Eq. (5.34)], the relaxation function [Eq. (5.40)], the density of states [Eq. (5.51)], the stretched exponent [Eq. (5.58)], and the spatial-dependent local diffusivity [Eq. (5.59)] later.

By using Eq. (5.28), Eq. (5.27) is transformed into

$$\left(\frac{\partial}{\partial t} - D_\nu q_\nu^2\right)\delta n(q_\nu, t) = 0,$$ (5.29)

where q_ν is the wavenumber on the fractal lattice. Eq. (5.29) reveals that the diffusivity D_ν is a constant. Therefore, holes should exhibit the Gaussian characteristic on the fractal lattice. The self-similarity of the fractal has dilation symmetry shown in Eq. (5.28). Taking the Fourier–Laplace transformation in time

$$\delta n\,[q_\nu, \omega] = \int_0^\infty \delta n(q_\nu, t)\exp(i\omega t)\,dt,$$

we obtain from Eq. (5.29)

$$\delta[\omega] = \sum_{q_\nu} \frac{\delta(q_\nu, t = 0)}{D_\nu q_\nu^2 - i\omega}.$$ (5.30)

It solves the initial value problem and is normalized by the number of lattice sites N. Note that Eq. (5.30) is valid only for small wavenumbers or a long wavelength. The local relaxation time τ_ν corresponding to the wavenumber q_ν is defined by

$$\tau_\nu = \frac{1}{D_\nu q_\nu^2}.$$ (5.31)

The Fourier–Laplace inversion of Eq. (5.30) is

$$\delta(t) = \sum_{q_\nu} \delta(q_\nu, t = 0)\exp\left[-\frac{t}{\tau_\nu(q_\nu)}\right] \equiv \sum_{q_\nu} \delta(q_\nu, t),$$ (5.32)

where the summation is carried out over the wavenumbers up to a cutoff q_c. Eq. (5.32) reveals that $\delta(q_v, t)$ is the solution of

$$\frac{df_i}{dt} = \frac{d\delta_i(t)}{dt} = -\frac{\delta_i(t)}{\tau_i}, \tag{5.33}$$

where the subscript i is identified with a particular wavenumber on the fractal lattice. Because the index i has the meaning of a wave vector, Eq. (5.33) is the discrete version of a continuous equation.

When the system is also under a temperature change, such as cooling, Eq. (5.33) has to be modified by using

$$\frac{d\bar{f}_i}{dt} = q\frac{\partial \bar{f}_i}{\partial T} = q\frac{\varepsilon_i \bar{f}_i}{kT^2}, \qquad i = 1, \dots, L,$$

where $q = dT/dt < 0$ is the cooling rate and ε_i represents the ith hole energy state. In addition, $\bar{f} = \sum \bar{f}_i$, $\varepsilon = \sum \varepsilon_i \bar{f}_i/\bar{f}$, and L is an integer that corresponds to the maximum wavenumber q_c. Combining Eq. (5.33) and the above equation, we obtain

$$\frac{d\delta_i(t)}{dt} = -\frac{\delta_i(t)}{\tau_i} - q\frac{\varepsilon_i \bar{f}_i}{kT^2}, \qquad i = 1, \dots, L. \tag{5.34}$$

This equation has the discrete form of the well-known KAHR (Kovacs–Aklonis–Hutchinson–Ramos) equation [9], which was presented as a phenomenological equation. It has been frequently used in the study of the thermal history behavior of glassy polymers; however, its main difficulty has been the proper choosing of the many adjusting parameters. On the other hand, Eq. (5.34) is derived and expressed by microscopic parameters that can be determined independently, as we shall continue to see in chapters 5 and 6. When the distribution of relaxation times is a delta function ($L = 1$), we will have a single kinetic equation that may be adequate to explain the structural relaxation phenomenon but fail to describe completely the memory effect (see Section 5.6). Stochastically, this failure is because of the non-Markovian nature of Eq. (5.34) with $L \neq 1$.

5.4 Relaxation Function and Time

For a system started from equilibrium, the solution of Eq. (5.34) is

$$\delta(T, t) = \sum_i \delta_i = -\frac{\varepsilon}{k} \int_0^t \frac{q\bar{f}}{T^2} \Phi(t - t')\, dt', \tag{5.35}$$

where Φ is the relaxation function. The $\Phi(t)$ here is basically a sum of an exponential, so that naturally a stretched exponential is present. Different paths of integration describe different thermal history behavior of the glassy-state relaxation and recovery kinetics. Typical examples of the thermal history paths for the above equation will be illustrated in Section 5.6.

In accordance with the anomalous diffusion on fractal lattice, one expects

$$\langle \Delta r^2 \rangle^{1/2} \equiv R \sim t^\beta, \qquad 0 < \beta \leq 1. \tag{5.36}$$

When the random process is Gaussian, we have $\beta = 1/2$ [see Eq. (2.12)]. As we mentioned earlier, regions of size with length scale l are separated by the energy barriers in an amorphous solid. The size is proportional to the time scale λ needed for a hole to penetrate a barrier of height ΔA, called the local activation energy [10]:

$$l \sim \lambda \sim \exp\left(\frac{\Delta A}{kT}\right). \tag{5.37}$$

Hence,

$$\frac{R}{l} \sim \frac{t^\beta}{\lambda} = \left(\frac{t}{a}\right)^\beta, \qquad \text{with} \quad a \sim \lambda^{1/\beta}, \tag{5.38}$$

where a is called the shift factor. It is the macroscopic relaxation time scale mentioned in Figure 1.2. During isothermal annealing, the frozen-in structure of a quenched glass starts to relax and the relaxation function $\Phi(t)$ can be interpreted as the probability that holes have not reached their equilibrium states. The probability of a hole, in the ith wavenumber state, having reached equilibrium in a time interval t is $(n_i/n)(R/l) \approx R/Ll$, where L is shown in Eq. (5.34). Thus, we write

$$\Phi(t) \approx \left(1 - \frac{R}{Ll}\right)^L \rightarrow \exp\left(-\frac{R}{l}\right), \qquad \text{as} \quad L \rightarrow \infty, \tag{5.39}$$

which because of the macroscopic spatial homogeneity is independent of i. Combining the above equation with Eq. (5.38) yields

$$\Phi(t) = \exp\left[-\left(\frac{t}{\tau}\right)^\beta\right], \qquad 0 < \beta \leq 1, \tag{5.40}$$

where $\tau \equiv \tau_r a$ is the macroscopic relaxation time and τ_r is a constant relaxation time at temperature T_r near T_g. The stretched exponential, Eq. (5.40), has the familiar form of the KWW (Kohlrausch–Williams–Watts) equation [11,12], except that β and τ are not empirical constants in the present theory, and they will be discussed later.

In the present lattice model, each lattice site occupies a single cell of volume v. In view of the cooperative nature of the hole motion, the barrier energy introduced in Eq. (5.37) is treated by a mean field average and related the Gibbs free energy per molecule (Δg) in a system restrained to single occupancy of cells by

$$\Delta A = \frac{N}{n}\Delta g = \frac{N}{n}(\varepsilon + pv - T\Delta s), \tag{5.41}$$

where the pressure p is chosen to be zero in the present discussion and Δs is the configuration entropy per molecule. Because the physical properties of glasses vary slowly in time, the system is in a quasi-equilibrium state. The entropy from the hole motion can be written approximately as

$$\Delta s = k \ln \left[\frac{n(t)}{\bar{n}} \right] = k \ln \left[1 + \frac{\delta(t)}{\bar{f}} \right]. \tag{5.42}$$

Combining eqs. (5.2), (5.37), (5.41), and (5.42) yields

$$a(T, \delta) = \left(\frac{\bar{f} + \delta}{f_r} \right)^{-1/[\beta(\bar{f} + \delta)]}. \tag{5.43}$$

At $T = T_r$, we have $\delta = 0$, $\bar{f} = f_r$, and $\tau = \tau_r$, which require $a = 1$. In practice, the shift factor is measured on the logarithmic scale, and Eq. (5.43) becomes

$$\ln a(T, \delta) = -\frac{1}{\beta(\bar{f} + \delta)} \ln \left(\frac{\bar{f} + \delta}{f_r} \right)$$

$$= \frac{1}{\beta(\bar{f} + \delta)} \left[\left(1 - \frac{\bar{f} + \delta}{f_r} \right) + \frac{1}{2} \left(1 - \frac{\bar{f} + \delta}{f_r} \right)^2 + \cdots \right],$$

$$\text{for} \quad \left| 1 - \frac{\bar{f} + \delta}{f_r} \right| < 1.$$

Therefore,

$$\ln a(T, \delta) = -\frac{1}{\beta} \left(\frac{1}{\bar{f} + \delta} - \frac{1}{f_r} \right), \quad \text{as} \quad \frac{\bar{f} + \delta}{f_r} \to 1. \tag{5.44}$$

The presence of δ in the above equation is caused by the change in the configurational hole entropy of nonequilibrium glasses.

For temperature above the glass transition, we have $\delta = 0$, and Eq. (5.44) reduces to the exact form of the Doolittle equation [13], which can also be written in the form of the well-known WLF (Williams–Landel–Ferry) equation [14] as

$$\log a = -\frac{C_1(T - T_r)}{C_2 + (T - T_r)}, \quad \text{for} \quad T > T_r, \tag{5.45}$$

where $C_1 = 1/2.303\beta f_r$, $C_2 = f_r/\Delta\alpha$, and $\Delta\alpha$ is given by Eq. (5.6). For temperature not too far below the glass transition, it is more convenient to have Eq. (5.44) expressed in the form

$$\ln a(T, \delta) = -\frac{\Delta\alpha(T - T_r) + \delta}{\beta f_r^2}. \tag{5.46}$$

This equation is going to be used in the quantitative prediction of volume relaxation and recovery (see Section 5.6), viscoelastic relaxation, and glassy-state deformation (see Chapter 6). Experimental verification of the theory will also be illustrated.

5.5   Relaxation Spectrum

The stretched exponent β resulted from the analysis of the last section should be a function of the local interaction that is related to the exponent ν mentioned in eqs. (5.27) and (5.28). Let us introduce the hole density–density correlation function

$$C(\vec{r}, t) = \frac{\langle \delta n(\vec{r}, t)\delta n(\vec{0}, 0)\rangle}{\langle \delta n^2 \rangle}. \tag{5.47}$$

The angular brackets denote an equilibrium ensemble average. $C(\vec{r}, t)$ is invariant under translations of $\vec{r}$ and t. It can be defined as the Green's function that vanishes when $\vec{r}$ or t are very large. We seek the solution of the equation

$$\left(\frac{\partial}{\partial t} - \vec{\nabla} \cdot D\vec{\nabla}\right) C(\vec{r}, t) = \delta(\vec{r})\delta(t), \tag{5.48}$$

where δ is Dirac's delta function. We follow the same procedure that led to Eq. (5.30), except we use a two-sided Fourier time transform for Eq. (5.48). To include all mode wavenumbers, we take the superposition of solutions

$$C(\omega) = \sum_{q_\nu} \frac{1}{D_\nu q_\nu^2 - i\omega} = \int_0^{q_c} \frac{\rho(q_\nu)\, dq_\nu}{D_\nu q_\nu^2 - i\omega}, \tag{5.49}$$

where ρ is the density of states. The number of modes per unit length along the hole path with the wavenumber between Q and $Q + dQ$ can be expressed by the number of modes on the fractal lattice by using Eq. (5.28)

$$\frac{dQ}{2\pi} \sim \frac{d_h}{2\pi} q_\nu^{d_h - 1}\, dq_\nu. \tag{5.50}$$

Using Eq. (5.31), we obtain the density of state

$$\rho(q_\nu)\, dq_\nu \sim q_\nu^{d_h - 1}\, dq_\nu \sim \tau_\nu^{-(1 + d_h/2)}\, d\tau_\nu. \tag{5.51}$$

Its time dependence is diffusive in nature. Substituting Eq. (5.51) into Eq. (5.50) leads to the asymptotic solution

$$C(\omega) \sim \int_\tau^\infty \frac{\tau_\nu^{-d_h/2}\, d\tau_\nu}{1 - i\omega\tau_\nu} \approx \int_\tau^\infty \frac{\tau_\nu^{-d_h/2}\, d\tau_\nu}{-i\omega\tau_\nu} = \frac{2}{d_h}\frac{\tau^{-d_h/2}}{i\omega}, \quad \text{for} \quad \omega\tau_\nu \gg 1, \tag{5.52}$$

where

$$\tau = \frac{1}{D_\nu q_c^2} \equiv \tau_r a \tag{5.53}$$

is the macroscopic relaxation time mentioned in Eq. (5.40).

The change in the state of glass during isothermal annealing is accompanied by dissipation of energy that is related to the hole density fluctuations. In accordance with the method of the generalized susceptibility (see Chapter 3), the viscoelastic loss modulus E'', which measures the energy dissipation, can be determined from Eq. (5.52) as

$$E'' \sim \operatorname{Im} C(\omega) \sim \tau^{-d_h/2}. \tag{5.54}$$

As mentioned in chapters 3 and 4, the viscoelastic relaxation modulus can in general be written in the form

$$E(t) = E_\infty + (E_0 - E_\infty)\Phi(t), \tag{5.55}$$

where E_0 and E_∞ are unrelaxed and relaxed moduli, respectively. The form of this equation is also valid for different physical quantities, like the bulk, shear, or tensile moduli. The stretched exponent β and the shift factor a are independent of the type of stress field applied to the system (see Section 5.6 and Chapter 6).

Substitution of Eq. (5.40) into the above equation gives the loss modulus [see Eq. (6.24)]

$$\frac{E''}{E_0 - E_\infty} = \sum_{m=1}^{\infty} \frac{(-1)^{m+1}\Gamma(m\beta + 1)}{m!(\omega\tau)^{m\beta}} \sin(m\beta\pi/2), \tag{5.56}$$

where Γ is the gamma function. The leading term provides a useful asymptotic expression in the glassy state

$$E'' \sim (\omega\tau)^{-\beta}, \quad \text{for} \quad \omega\tau \gg 1. \tag{5.57}$$

Comparing Eq. (5.54) and (5.57) yields

$$\beta = \frac{d_h}{2} = \frac{1}{2+\nu}. \tag{5.58}$$

When we look at eqs. (5.51) and (5.52), Eq. (58) confirms the way of relating the relaxation function, Eq. (5.40), to the relaxation time spectrum by the stretched exponent β. The glassy-state relaxation is dominated by the part of the spectrum having longer relaxation times. Parallel to Eq. (5.28), Eq. (5.38) suggests another self-similar scaling between the local and global relaxation time scales $a \sim \lambda^{2/d_h}$. This extends the use of dilation symmetry from space to time in polymeric glasses.

The fractal dynamics of holes are diffusive, and the diffusivity depends strongly on the tenuous structure in fractal lattices. Using eqs. (5.36) and (5.58), we obtain the dependence of the local diffusion coefficient (D) on diffusion length (R) as

$$D = R^2/2t \sim R^{-\nu}. \tag{5.59}$$

The divergence of the diffusivity in the above equation at $R = 0$ for $\nu > 0$ was the main reason behind the fractal dimension that was introduced in Eq. (5.27) together with a spatial transformation, Eq. (5.28). For linear polymers, $d_h = 1$, eqs. (5.58) and (5.59) show that the diffusion coefficient is spatially independent because the spreading of excess holes is Gaussian ($\nu = 0$). The value of $\beta = 1/2$ is found to be in agreement with experimental data for linear polymers (see Section 5.6). When $0 < d_h < 1$, we have $\beta < 1/2$ and $\nu > 0$; the motion of holes slows because the local diffusion coefficient in Euclidean space decreases with distance. In the study of crosslinked polymers, the interesting range is $\nu \geq 0$ and the value of ν usually increases with the crosslink density. The mass (M)-size (L) scaling for chain networks is described by the static Hausdorff dimension (d_f): $M \sim L^{d_f}$ (see Appendix 4A). We find that d_f is related to ν by $d_f = 2(\nu + 2)/3$. For an ideal phantom network, we have $\nu = 4$ that leads to $d_h = 1/3$ and $d_f = 4$ (see Section 4.10), which has been verified experimentally (see Table 4.1).

5.6   Volume Relaxation and Recovery

We have analyzed the hole dynamics and fluctuations on a fractal lattice and have obtained the solution for the nonequilibrium hole fraction for a system started from equilibrium, as in Eq. (5.35):

$$\delta(t) = \frac{V(t) - \overline{V}}{\overline{V}} = -\frac{\varepsilon}{k} \int_0^t \frac{q\bar{f}}{T^2} \Phi(t, t') \, dt'.$$

Different paths of time integration in this equation describe different thermal history behavior of the glassy-state relaxation and the memory effect of volume recovery. In the case of volume relaxation during isothermal annealing followed by quenching from an equilibrium temperature T_0 to the annealing temperature T_1 (i.e., $q_0 \to -\infty$ and $q_1 = 0$, shown in Figure 5.4, Eq. (5.35) becomes

$$\delta(t) = -\Phi(t) \int_{T_0}^{T_1} \frac{\varepsilon \bar{f}(T')}{kT'^2} \, dT' = [\bar{f}(T_0) - \bar{f}(T_1)]\Phi(t). \qquad (5.60)$$

The isothermal volume relaxation of glass-forming polymers following a single temperature jump from equilibrium can be calculated from eqs. (5.2), (5.40), (5.46), and (5.60). Figure 5.4 compares the calculated and measured volume contractions for PVAc after quenches from equilibrium at $T_0 = 313$ K to several different annealing temperatures T_1 ranging from 298 K to 310.5 K. By using the predetermined ε and f_r mentioned in Section 5.1, the parameters $\beta = 0.48$ and $\tau_r = 25$ min were adopted to fit experimental data.

The effective relaxation time, $\tau_{eff} = [-d \ln \delta / dt]^{-1}$, helps us to see the asymmetric characteristics between the contraction and expansion isotherms quenched and rapid heated, respectively, from different initial equilibrium temperatures T_0 to T_1. It is a more severe test of any theoretical treatment. Using eqs. (5.40) and (5.60), we obtain

$$\log \tau_{eff} = \log \frac{\tau}{\beta} + \frac{1 - \beta}{\beta} \log \left[-\ln \Phi(t) \right]. \qquad (5.61)$$

A comparison of Eq. (5.61) and experimental data is shown in Figure 5.5. Five different initial temperatures (T_0) exist with a common annealing temperature

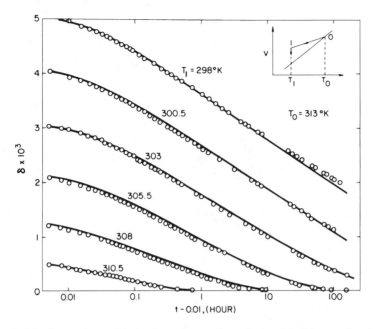

FIGURE 5.4. Comparison of the isothermal annealing calculated (solid curves) and measured (circles [2]) for PVAc quenched from T_0 to various T_1.

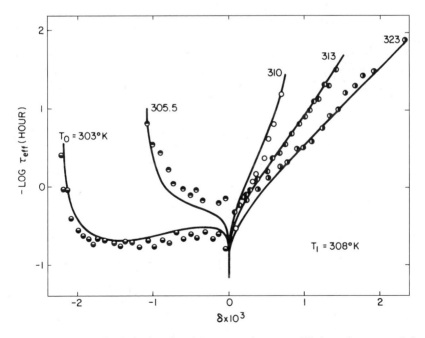

FIGURE 5.5. The effective relaxation time versus the nonequilibrium glassy state δ for PVAc. A comparison between theory (solid curves) and experiment (circles [2]).

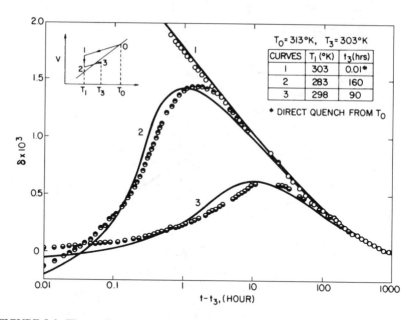

FIGURE 5.6. The predicted (solid curves) and measured (circles [2]) memory effects are compared for PVAc.

$T_1 = 308$ K. The asymmetry in approaching equilibrium from $\delta > 0$ and $\delta < 0$ is known as one of the most characteristic features of structure relaxation in polymer glasses.

The memory effect is associated with two consecutive temperature jumps. The thermal history of the system involves quenching $(0 \to 1)$, annealing $(1 \to 2)$, and rapid rehearing $(2 \to 3)$; i.e., $q_0 \to -\infty$, $q_1 \to 0$, and $q_2 \to \infty$, shown in Figure 5.6. The volume recovery is calculated as a function of elapsed time $t - t_3$ from Eq. (5.35) as

$$\delta(t) = -\frac{\varepsilon}{k} \left[\int_0^{t_1} \frac{q_0 \bar{f}}{T^2} \Phi(t, t') \, dt' + \int_{t_2}^{t_3} \frac{q_2 \bar{f}}{T^2} \Phi(t, t') \, dt' \right]$$

$$= [\bar{f}(T_0) - \bar{f}(T_1)]\Phi_1(t) - [\bar{f}(T_3) - \bar{f}(T_1)]\Phi_3(t), \quad \text{for } t > t_3. \quad (5.62)$$

Here,

$$\Phi_1(t) = \exp \left\{ - \left[(\ln \Phi(t_2 - t_1))^{1/\beta} + \frac{t - t_3}{\tau(T_3, \delta)} \right]^\beta \right\}$$

and

$$\Phi_3(t) = \exp \left\{ - \left[\frac{t - t_3}{\tau(T_3, \delta)} \right]^\beta \right\}.$$

The input parameters for PVAc with $T_r = 308$ K are again to be

$$\varepsilon = 2.51 \, \text{kcal/mol}, \quad f_r = 0.0336, \quad \beta = 0.48, \quad \tau_r = 25 \, \text{min}. \quad (5.63)$$

A comparison between the theoretical and experimental behavior is shown in Figure 5.6. This plot is a much more severe test of the theory than the single temperature jump. Curve 1 is the intrinsic isotherm obtained by direct quenching from $T_o = 313$ K to $T_1 = 303$ K, already shown in Figure 5.4. Interestingly, figures 5.4 to 5.6 reveal that all of the salient features of the volume relaxation and recovery in glass-forming polymers can be described by the theory with the same set of input parameters given by Eq. (5.63). The distribution of relaxation time expressed by β is independent of temperature. Of course, all of the memory effects shown in Figure 5.6 would disappear if $\beta = 1$, which can be seen from Eq. (5.62): $\delta = 0$ for $\beta = 1$.

5.7   PVT Equation of State

The PVT behavior of amorphous polymers in both the equilibrium liquid and nonequilibrium glassy states is another fundamental aspect that is important to the understanding of material properties. When an amorphous melt is cooled isobarically from liquid to glass, the process of forming glassy polymers involves the slow relaxation of a frozen-in structure. By converting time to temperature, the nonequilibrium state given by Eq. (5.35) can be rewritten in the form

$$\delta(t) = -\frac{\varepsilon + pv}{k} \int_{T_0}^{T} \frac{\bar{f}}{T'^2} \exp\left[-\left(\frac{T - T'}{|q|\tau} \right)^\beta \right] dT', \quad (5.64)$$

where T_0 is the initial temperature in the equilibrium liquid state above T_g, and τ is given by Eq. (5.46). The total volume defined by Eq. (5.4) is

$$V(T, p, q) = xn_x v(T, p) \frac{1 + \delta(T, p, q)}{1 - \bar{f}(T, p)} \equiv \overline{V}(T, p)\,[1 + \delta(T, p, \delta)]. \quad (5.65)$$

The occupied volume $xn_x v$ is independent of T_g, but it is allowed to expand and contraction linearly,

$$v(T, p) = v_r \,[1 + \alpha_v(T - T_r) - \kappa_v p], \quad (5.66)$$

where α_v was mentioned in Eq. (5.5) and κ_v is the isothermal compressibility.

In the case of $p = 0$, the calculated equilibrium and nonequilibrium volumes of PVAc are compared with experimental data in Figure 5.7. The input parameters include Eq. (5.63), and the parameters that describe the temperature dependence of the occupied volume, $xn_x v_r$, α_v, and κ_v, which had already been determined from Figure 5.1. The numerical solution of eqs. (5.64)–(5.66) is not affected by the choice of the initial temperature as long as $T_0 \geq T_g + 10$ K. The effect of cooling

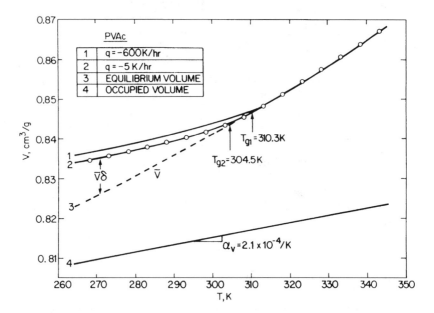

FIGURE 5.7. The equilibrium and nonequilibrium volumes (V) of PVAc are plotted as a function of temperature (T) and cooling rate (q). Circles are experimental data [15]. The lattice thermal expansion coefficient (α_V) is related to the occupied volume and has nothing to do with the glass transition temperature (T_g). The kinetics of the glass transition is determined by the dynamics of hole motion.

rate is also calculated in the Figure 5.7. The difference in the glass-transition temperatures can be adequately accounted by (see Chapter 6)

$$T_{g1} - T_{g2} = \frac{\beta f_r^2}{\Delta \alpha} \ln \left(\frac{q_1}{q_2} \right) = 2.79 \log \left(\frac{q_1}{q_2} \right), \qquad (5.67)$$

where $\Delta \alpha$ is given by Eq. (5.6) as the excess thermal expansion coefficient of the liquid with respect to the glass. The calculation reveals that the glass transition is a kinetic phenomenon.

When a system is under pressure with negligible pressure rate, the PVT behavior is again calculated from eqs. (5.2), (5.46), (5.64), (5.65), and (5.66). A comparison of the calculated PVT behavior and the isobaric volume data on PVAc for temperature from 250–350 K and pressure up to 800 bar is shown in Figure 5.8. The cooling rate is 5 K/hr. In addition to the same set of input free volume and lattice parameters used in the calculations for Figure 5.7, $\kappa_V = 3.4 \times 10^{-5}$/bar is needed in the calculation of Figure 5.8. The compressibility can be determined independently under atmospheric pressure [16; also see Chapter 6]. Pressure has a strong effect on T_g. Again, the nonequilibrium state δ and its relaxation have served as the foundation for the quantitative description of the PVT behavior and the glass-transition temperature.

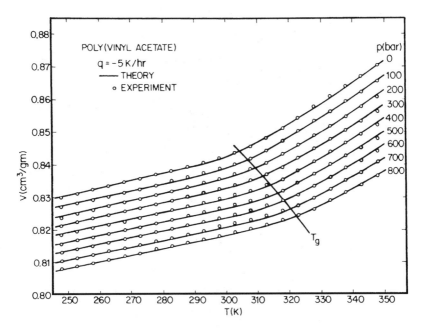

FIGURE 5.8. Comparison of the calculated (solid curves) and measured (circles [15]) PVT behavior of PVAc. It also reveals the effect of pressure on T_g.

References

1. R. H. Howard (Ed.), *Physics of Glassy Polymers* (Wiley, New York, 1973).
2. A. J. Kovacs, Adv. Poly. Sci. **3**, 397 (1963).
3. T. S. Chow, Macromolecules **17**, 2336 (1984).
4. T. S. Chow, Macromol. Theory Simul. **4**, 397 (1995).
5. Q. Deng, F. Zandiehnadem, and Y. C. Jean, Macromolecules **25**, 1090 (1992).
6. T. S. Chow, Macromolecules **22**, 701 (1989).
7. T. S. Chow, Phys. Rev. A **44**, 6916 (1991).
8. B. B. Mandelbrot, *The Fractal Geometry of Nature* (Freeman, San Francisco, 1982).
9. A. J. Kovacs, J. J. Aklonis, J. M. Hutchinson, and A. R. Ramos, J. Polym. Sci., Polym. Phys. Ed. **17**, 1097 (1979).
10. S. K. Ma, *Statistical Mechanics* (World Scientific, Philadelphia, 1985).
11. R. Kohlrausch, Ann. Phys. **21**, 393 (1847).
12. G. Williams and D. C. Watts, Trans. Faraday Soc. **66**, 80 (1970).
13. A. K. Doolittle, J. Appl. Phys. **22**, 1471 (1951).
14. M. L. Williams, R. F. Landel, and J. D. Ferry, J. Am. Chem. Soc. **77**, 2701 (1955).
15. J. E. McKinney and M. Goldstein, J. Res. Nat. Bur. Standards **78**A, 331 (1974).
16. T. S. Chow, J. Rheol. **30**, 729 (1986).

6

Glassy Polymers

The physical properties of glassy polymers vary more strongly with time and temperature than those of other materials, such as metal and ceramics, because of the inherent, irreversible nature of glassy polymers and their significantly lower glass transition temperature. The purpose of this chapter is trying to establish a direct link between the structural relaxation and the mechanical properties of polymers in the glassy state. Because the nonequilibrium structure relaxes very slowly in the glassy state, the long time behavior and thermal history become a major concern. This characteristic is usually known as the physical aging phenomena, and the discussions here are going to be mainly devoted to the volumetric and mechanical properties.

Glasses are characterized by the absence of long range order. In the last chapter, the glassy-state relaxation was analyzed as a result of the local configurational rearrangements of molecular segments, which is quantified by the hole motion. Following the concept of hole dynamics, pertinent equations will be derived (1) for the elucidation of the kinetics of the glass transition, (2) for the calculation of viscoelastic response, and (3) for the prediction of plastic yield and nonlinear stress–strain behavior. During the processes, two useful concepts will be introduced and analyzed. The first one is the physical aging exponent that characterizes the thermal history behavior in the vicinity of the glass transition. At high stress levels, the activation volume tensor is the other one that plays a key role in the nonlinear viscoelasticity.

6.1 Glass Transition

The glass transition temperature is perhaps the most important physical parameter that one needs to decide on the application of the many amorphous polymers that are available today. This transition is evidenced by a change in the slope of a plot of specific volume versus temperature, as shown in Figure 5.7. Changes in the mechanical, and other, properties will also occur at the same temperature. This behavior results from the abrupt onset of extensive molecular motion as the temperature is raised or the freezing of such motion as the temperature is lowered. These motions are inhibited in the glassy state, in which the viscosity is so high that the specific volume cannot reach its equilibrium quickly in a practical time scale (see Figure 5.4).

Little doubt exists at this point that the glassy state represents a situation of frozen-in disorder. Because of its quasi-equilibrium state, a confusion occurs about the applicability of the laws of thermodynamics to the system, on the one hand, and the need for describing the observed kinetic phenomena, on the other hand. Nevertheless, two different views have been presented in the literature in the molecular interpretations of the glass transition. One view considers conditions when relaxation processes occur so slowly that the glass transition may be treated as a time-independent phenomenon. The other view is directed at the nonequilibrium character of structural relaxation and physical aging. We shall discuss both approaches.

6.1.A   Equilibrium Approach

Perhaps the most familiar example of a thermodynamic theory for polymers is that from Gibbs and DiMarzio (GD) [1]. The glass transition temperature (T_g) is assumed by GD to be a second-order phase transition temperature (T_2), at which the configurational entropy of the system is zero (see Section 7.7) and the number of conformations available to a macromolecule becomes less. Consequently, the molecule appears stiffer and the motion slower as T_2 is approached. In spite of (1) the problematic assumption of the second-order phase transition (see Figure 6.1) and (2) that it is impossible to reach T_2 in a practical sense, one may draw a range of conclusion in regard to the glass transition temperature from this thermodynamic theory. Within this framework, it is also assumed that those factors influencing T_2 are going to affect T_g in a similar fashion. This process has resulted in considerable success in describing the effects of molecular weight, plasticization, and crosslinking on the glass transition.

The GD theory uses the complex Flory combinatorial statistics [2] to calculate the configurational partition function by hindered rotation about the bonds of the main chain in the molecule. The theory takes into account both the intramolecular bond rotational energy (ε_f) and the intermolecular energy of hole formation (ε). The flex energy ε_f measures the stiffness and is the energy difference between the high-energy gauche and the low-energy trans states. Figure 6.1 shows that the glass transition is completely different from the second-order phase transition postulated in the GD theory. Therefore, it is desirable to remove the second-order phase transition requirement in the GD theory but to retain other essential features in a different approach.

The "equilibrium" glass temperature (T_r) may be treated as a thermodynamic anomaly, at which the most stable hole configuration is reached under the close

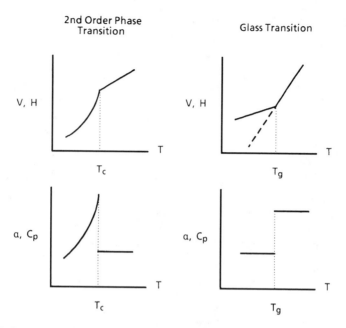

FIGURE 6.1. The second-order phase transition versus the glass transition. They have different temperature dependence of the volume (V) and enthalpy (H), and their derivatives α (thermal expansion) and C_p (specific heat).

packing of hole and flex bonds, rather than a phase transition. This anomaly enables a prediction of T_r to be made by $\varepsilon_f, \varepsilon$, and the degree of polymerization of a polymer [x in Eq. (5.1)] of a given coordinate number (z). Consider high molecular-weight linear polymers ($x \gg 100$), and choose $\bar{f}_r = 1/30$ and $z = 6$. A crude approximate expression for T_r is obtained [3]

$$\frac{\varepsilon}{kT_r} \cong \frac{2.15\varepsilon_f}{kT_r} = 4.16, \tag{6.1}$$

which has been experimentally verified for linear polymers, like PVAc, Poly(vinyl chlorode) (PVC), and Polystyrene (PS). Eq. (6.1) reveals that the ratio of hole and bond rotational energies is close to a constant, which supports the notion that the conformational theory is experimentally equivalent to the hole theory in the molecular interpretation of the glass transition.

6.1.B Nonequilibrium Approach

The purpose here is to derive a nonequilibrium criterion for the direct determination of the kinetic effects on the glass transition temperature. We would like to see the dependence of T_g on relaxation times. When an amorphous polymer is cooled from an increased equilibrium liquid temperature (see Figure 5.7), its physical properties undergo rapid change over the glass transition region. Making use of the basic equation (5.34), it is convenient to introduce an average relaxation time $\langle\tau\rangle$ defined by

$$\langle\tau\rangle^{-1} = \sum_i (\delta_i/\tau_i) \bigg/ \sum_i \delta_i. \tag{6.2}$$

In the glass transition region, Eq. (5.34) can be written in the form

$$\frac{d\delta}{dt} = -\frac{\delta}{\langle\tau\rangle} - q\Delta\alpha, \tag{6.3}$$

where the excess thermal expansion $\Delta\alpha$ is a constant given by Eq. (5.6) with $p = 0$.

The fictive temperature T_f, defined by $\delta = \Delta\alpha(T_f - T)$, has been reported in the literature [4,5] as another parameter that is useful in the investigations of vitrification. Eq. (6.3) is then rewritten for a system cooling from an equilibrium liquid state as

$$T_f = T + |q|\langle\tau\rangle\frac{dT_f}{dT}. \tag{6.4}$$

On cooling, the fictive temperature decreases linearly with temperature along the equilibrium line up to a transition region, where the change of slope occurs as a result of the frozen-in structure, after which the curve approaches its asymptote. Assume that T_g is near equilibrium ($dT_f/dT = 1$) at the entrance of the transition region. Eq. (6.4) is then reduced to $T_f \approx T + |q|\langle\tau\rangle$. The glass transition temperature is conventionally defined as the point at which the asymptotic line from the glassy state intersects the equilibrium liquid line. Thus, T_g is determined by using the freezing-in condition of $dT_f/dT = 0$ [6]. When all of the higher order terms are neglected, we obtain

$$\left.\frac{d\langle\tau\rangle}{dT}\right|_{T=T_g} = -\frac{1}{|q|}, \quad \text{at } T = T_g. \tag{6.5}$$

This result is the nonequilibrium criterion for the direct determination of T_g from the average relaxation time. No restriction is placed on the specific functional form of $\langle\tau\rangle$ as long as it is a continuous, decreasing function of temperature. In practice, the average relaxation time may be treated as the macroscopic relaxation time; i.e., $\langle\tau\rangle = \tau_r a$. In the transition region, we may substitute Eq. (5.46) with $\delta \approx 0$ into Eq. (6.5) and get

$$T_g = T_r + \frac{\beta f_r^2}{\Delta\alpha}\ln\left(\frac{\Delta\alpha}{\beta f_r^2}|q|\tau_r\right). \tag{6.6}$$

This expression leads to Eq. (5.67) in its interpretation of the effect of cooling rate on the PVT behavior of PVAC shown in Figure 5.7.

6.2   Physical Aging

When polymer is cooled from liquid to glass, the dependence of the macroscopic relaxation time scale on temperature is calculated from eqs. (5.35) and (5.46) and is shown in Figure 6.2 for PVAc. To evaluate the thermal history dependent integral, Eq. (5.35), dt' is replaced by dT'/q, $\bar{f}/T^2$ is treated as a function of T', and $t - t' = (T - T')/q$ in the cooling step [see Eq. (5.64)]. The departure from a Doolittle–WLF type of temperature dependence [see Eq. (5.45)] in the liquid state to an Eq. (5.46) type of shift factor (also see Figure 1.2) depends not only on temperature, but also on the thermal history of glasses. Cooling rate affects the magnitude but not the slope of $\log a$ for $T < T_r$. The change in the slope at temperatures below and above the glass transition can be expressed by the Arrhenius type of activation energy by following Eq. (5.46)

$$\Delta H = k \left.\frac{\partial(\ln a)}{\partial(1/T)}\right|_{T \to T_r} = -kT_r^2 \frac{\partial \ln a}{\partial T} = \frac{\varepsilon}{\beta f_r}\left(1 + \frac{1}{\Delta\alpha}\frac{\partial\delta}{\partial T}\right). \qquad (6.7)$$

The nonequilibrium transitions are originated from the second term in the above equation.

Struik has introduced an exponent (μ) in the glassy state to characterize the physical aging observed in his isothermal creep experiments [7],

$$a(T, t) \sim t^{\mu}, \qquad (6.8)$$

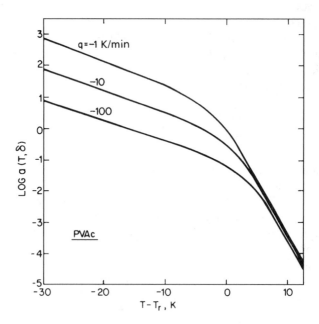

FIGURE 6.2. Calculated shift factor (a) as a function of temperature and cooling rate (q) as PVAc is vitrified through the glass transition region.

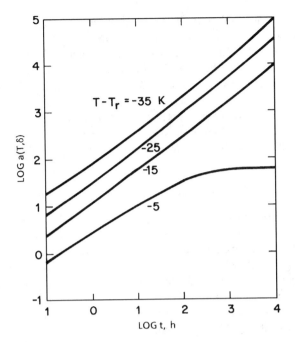

FIGURE 6.3. Calculated shift factor as a function of the annealing time (t) and temperature (T) of quenched PVAc in the vicinity of the glass transition region.

at the long aging time t. To calculate the aging exponent, let us consider the case of isothermal annealing in the glassy state followed by quenching from an increased liquid temperature. In addition to the cooling step mentioned earlier, the relaxation function in Eq. (5.35) is written the form, $\Phi(t-t') = \Phi[(T-T')/q\tau + (t-t_1)/\tau]$ in the isothermal annealing step, where $t_1 = (T - T_o)/q$ and $t - t'$ are ranges in logarithmic time scale. The dependence of the shift factor a on annealing (or physical aging) time of quenched [$q \to -\infty$, and see Eq. (5.60)] PVAc in accordance with Eq. (5.46) is shown in Figure 6.3. Clearly, the physical-aging exponent μ is a constant that is independent of annealing time and temperature in the glassy state.

In the vicinity of the glass transition, eqs. (5.67) and (6.6) suggest that

$$\beta f_r^2 d(\ln t) = (\Delta \alpha)\, dT. \tag{6.9}$$

This equation helps us to obtain the temperature dependence of the physical-aging exponent

$$\mu(T) = \frac{\partial(\log a)}{\partial(\log t)} = -\frac{1}{\beta f_r^2} \frac{\partial \delta}{\partial(\ln t)} = -\frac{1}{\Delta \alpha} \frac{\partial \delta}{\partial T}. \tag{6.10}$$

Substituting Eq. (6.10) into Eq. (6.7) gives

$$\Delta H(T) = \frac{\varepsilon}{\beta f_r}[1 - \mu(T)]. \tag{6.11}$$

This equation provides a simple relationship between the activation energy ΔH and the physical-aging exponent μ in the interpretation of the transitions of the relaxation time from the Doolittle–WLF type of dependence to the form of Eq. (5.46) near the glass transition shown in Figures 6.2. Figures 6.2 and 1.2 show that μ reaches a constant value of 0.8 for $T < T_r - 10\,\mathrm{K}$ and approaches zero for $T > T_g$.

The discussions carried out so far have been based on the *volume* (bulk) relaxation. The next step is to look at the transient properties of linear polymers under *shear* deformation. The shear relaxation modulus can in general be written as (see sections 4.6 and 5.5)

$$G(t) = G_\infty + (G_0 - G_\infty)\Phi(t), \tag{6.12}$$

where $\Phi(t)$ is the relaxation function given by Eq. (5.40). The relaxed modulus (G_∞) is always much smaller than the unrelaxed modulus (G_0) in the glassy state. A comparison between the theory and experiment for PVAc is made out in figures 6.4 and 6.5, where the master creep curve and its shift factor, respectively, are shown. Creep measures the slow deformational process under a fixed load in the linear viscoelastic range. Because of the long relaxation time in the glassy state, the creep compliance (J) may be determined from Eq. (6.12) in accordance with the quasi-elastic approximation: $J = 1/G$. The input parameters include Eq. (5.63), $G_0(T) = 1.2 \times 10^9 (297/T) Nm^{-2}$, and negligible G_∞. The creep data for $t/a > 10^5$ sec, which represent those measurements carried out above T_g, are lower than the calculated curve. This decrease suggests that the relaxed modulus G_∞ may no longer be negligible in comparison with G_0 in this

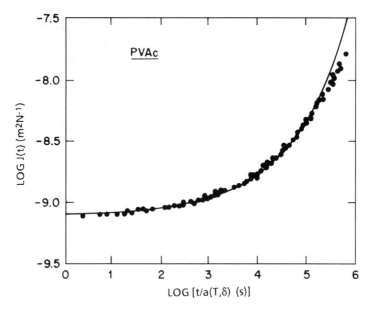

FIGURE 6.4. Master shear creep curve of PVAc. We see a comparison between the theory (solid curves) and experiment (points [8]).

TABLE 6.1. *Hole and activation energies for poly(vinyl acetate) [11]*

Energy	Formula	kcal/mol
Mean hole energy	ε	2.51
Local activation energy	ε/f_r	74.7
Activation energy for $T > T_g$	$\varepsilon/\beta f_r$	155.6
Activation energy for $T < T_g$	$\Delta H = (1-\mu)\varepsilon/\beta f_r$	30.5

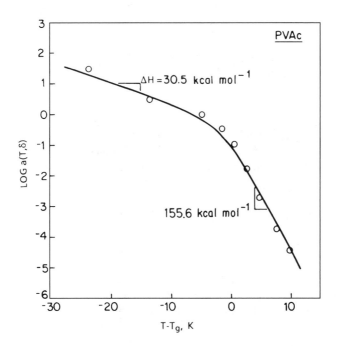

FIGURE 6.5. A comparison of the calculated (solid curves) and measured (circles [8]) shift factor of PVAc. The experimental data were obtained from the shear creep measurement.

particular situation. The prediction of Eq. (5.46), in which eqs. (5.2), (5.35), and (5.40) were used, is compared with the experimental $\log a - T$ data in Figure 6.5, where $\log a$ is chosen to be zero at $T = 297$ K. The change of the activation energies in this figure from $\varepsilon/\beta f_r = 155.6$ kcal/mol to $\varepsilon/\beta f_r(1-\mu) = 30.5$ kcal/mol in the vicinity of T_g is interpreted by the temperature-dependent physical-aging exponent. The quantitative agreement between Figures 6.2 and 6.5 suggests that both the bulk and shear deformations share the *same* relaxation mechanism. The relationships between the intermolecular hole energy and the different activation energies are summarized in Table 6.1.

In the glassy state, the relaxation modulus is often represented by a power law, $G(t) \sim t^{-m}$, with its exponent m determined from eqs. (6.12) and (5.40):

$$m = -\frac{d(\log G)}{d(\log t)}\bigg|_{t=\tau} = (1-\mu)\beta, \qquad (6.13)$$

which is known as the Alfrey approximation (see Section 4.11). At this point, we would like to point out the important differences between Eq. (5.40) and the KWW equation, which has the same form as Eq. (5.40), except that it replaces β and τ by β_w and τ_w, respectively. The β_w and τ_w are treated as empirical parameters that must alter continuously throughout the glass transition region to fit the experimental data. Although τ is a time-dependent quantity, β has already been experimentally verified to be a constant in sections 5.6, 5.7, and 6.2. By using the hole parameters, Eq. (5.63), for PVAc and Eq. (6.13), an explicit comparison between β_w and β can be made by looking at the relaxation function:

$$t^{\beta_w} \sim -\ln \Phi \sim t^{(1-\mu)\beta} \sim \begin{cases} t^{0.096}, & T < T_g - 10\,\text{K}, \\ t^{0.48}, & T > T_G. \end{cases} \tag{6.14}$$

It illustrates the drastic change of β_w from 0.48 to 0.096 for PVAc as it is cooled from liquid to glass in the vicinity of T_g. On the other hand, β remains to be a constant of 0.48 because the physical-aging exponent μ changes with temperature. The constant β is essential for the model that has the capability of making the quantitative predictions of the most salient features about the glassy-state relaxation and deformation. We shall continue to explore them in the rest of this chapter.

6.3 Dynamic Viscoelastic Properties

A successful glass theory should not be limited in its application to the transient bulk and shear properties of linear polymers; it has to also be capable of making molecular interpretation and prediction of the dynamic viscoelastic properties of crosslinked polymers. The constitutive equation of the tensile stress (σ) and strain (e) for linear viscoelastic solids can in general be written as

$$\sigma(t) = \int_{-\infty}^{t} E(t-s)\dot{e}(s)\,ds, \tag{6.15}$$

where E is the tensile relaxation modulus. By taking to the Fourier–Laplace transform and putting $\omega t = (\omega \tau)(t/\tau) \equiv zy$, Eq. (6.15) becomes [see Eq. (4.86)]

$$\sigma[z]/e[z] = E^*(z) = iz \int_0^{\infty} E(y)\exp(-izy)\,dy. \tag{6.16}$$

The complex tensile modulus is then separated into the real and imaginary parts: $E^*(z) = E'(z) + iE''(z)$. The real part is the storage modulus, and the imaginary part is the loss modulus, which defines the energy dissipation.

The general form of Eq. (6.12) should apply to the tensile modulus as well:

$$\frac{E(y) - E_\infty}{E_0 - E_\infty} = \Phi(y) = \exp(-y^\beta), \qquad 0 < \beta \leq 1, \tag{6.17}$$

where E_0 and E_∞ are the unrelaxed and relaxed moduli, respectively. Substitution of Eq. (6.17) into Eq. (6.16) yields

$$\frac{E' - E_\infty}{E_0 - E_\infty} = \frac{\omega\tau}{\beta} \int_0^\infty \exp(-x)\left[x^{(1-\beta)/\beta} \sin(\omega\tau x^{1/\beta})\right] dx \qquad (6.18)$$

and

$$\frac{E''}{E_0 - E_\infty} = \frac{\omega\tau}{\beta} \int_0^\infty \exp(-x)\left[x^{(1-\beta)/\beta} \cos(\omega\tau x^{1/\beta})\right] dx. \qquad (6.19)$$

These two integrals can be evaluated numerically for small $\omega\tau$.

When $\omega\tau$ is large, the convergence of the numerical integrations becomes so slow that an alternative approach has to be developed. Let us consider the Laplace transform by replacing $iz = p$ in Eq. (6.16). According to the mathematical theorem,

$$\lim_{p\to\infty} p\Phi(p) = \lim_{y\to 0} \Phi(y), \qquad (6.20)$$

we take the series expansion of Eq. (6.17)

$$\Phi(y) = \exp(-y^\beta) = 1 - y^\beta + \frac{y^{2\beta}}{2!} - \frac{y^{3\beta}}{3!} + \cdots. \qquad (6.21)$$

The Laplace transform of Eq. (6.21) gives

$$p\Phi(p) - 1 = \sum_{m=1}^\infty (-1)^m \Gamma(m\beta + 1)/m! p^{m\beta}, \qquad (6.22)$$

where Γ is the gamma function. Thus, for large $\omega\tau$, the storage and lost moduli can be efficiently calculated, respectively, from

$$\frac{E' - E_\infty}{E_0 - E_\infty} = 1 + \sum_{m=1}^\infty \frac{(-1)^m \Gamma(m\beta + 1)}{m! z^{m\beta}} \cos(m\beta\pi/2) \qquad (6.23)$$

and

$$\frac{E''}{E_0 - E_\infty} = \sum_{m=1}^\infty \frac{(-1)^{m+1} \Gamma(m\beta + 1)}{m! z^{m\beta}} \sin(m\beta\pi/2). \qquad (6.24)$$

The leading terms of the above two equations give a useful asymptotic expression for the loss tangent $(= E''/E')$ in the glassy state, where E_∞ is usually negligible. That is,

$$\lim_{\omega\tau\to\infty} \tan\Delta = (\omega\tau)^{-\beta}\Gamma(1 + \beta)[\sin(\beta\pi/2) + \cos(\beta\pi/2)]. \qquad (6.25)$$

When $\beta = 1$, the above equation becomes

$$\lim_{\omega\tau\to\infty} \tan\Delta = (\omega\tau)^{-1}, \qquad (6.26)$$

which has the familiar form.

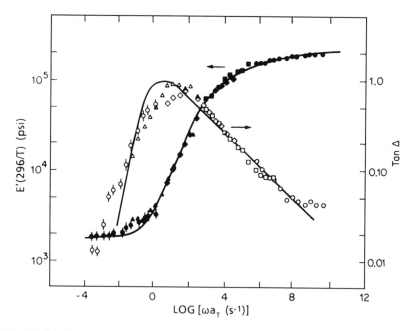

FIGURE 6.6. Comparison of the calculated (solid curves) and measured (points [9,10]) master curves of the dynamic viscoelastic properties of an epoxy resin.

A comparison between the theory and experiment for epoxy resins is carried out in figures 6.6 and 6.7. The dynamic viscoelastic data cover a four-decade frequency range from 0.01 to 100 Hz and temperatures from 25 to 200 °C. The experimental points in Figure 6.6 have been shifted horizontally to form the master curves. The reference temperature for the master curves is 165 °C. The full curves represent the theoretical calculation, where $(E_0, E_\infty) = (1.38, 1.24 \times 10^{-2})(296/T)$ Gpa. Using eqs. (6.18), (6.19), (6.23), and (6.24), we have determined $\beta = 0.19$, which defines the shape of the curves and the range of time scale for the master storage modulus and lost tangent. Because the abscissa in Figure 6.6 is expressed by ωa_T rather than the dimensionless $\omega\tau$, we are also able to determine $\tau(165\,°C) = 10^{-3.8}$ sec. Both figures 6.6 and 6.7 reveal that the time scale for crosslinked networks covers the range far broader than that for linear polymers.

When the experimental a_T versus T data on the epoxy resin above $T_r = 115\,°C$ ($\delta = 0$) are used, eqs. (5.2) and (5.46) give $\varepsilon = 4.5$ kcal/mol and $f_r = 0.13$. In analyzing the data, we consider $\log a_T = 8.58 + \log a$ with $a(115\,°C) = 1$ in Figure 6.7. The activation energy changes from $\varepsilon/\beta f_r = 182.2$ to 84 kcal/mol $= (1 - \mu)\varepsilon/\beta f_r$ as the epoxy is cooled through the glass transition region. The physical-aging exponent μ reaches a constant of 0.54 for $T < T_g - 10\,°C$ and approaches zero for $T > T_g$. The value of μ for crosslinked polymers is lower than that for linear polymers in the glassy state. Again, we see the transition from WLF dependence to an Arrhenius temperature dependence of the shift factor as a result of the critical slowing of relaxation processes, which has been explained by the

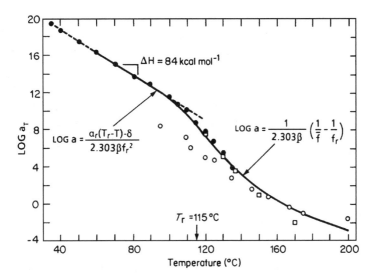

FIGURE 6.7. A comparison of the calculated (solid curves) and measured (points [9]) shift factor of an epoxy resin.

temperature-dependent physical-aging exponent. From eqs. (6.8) and (6.25), we have

$$\tan \Delta \sim (\omega \tau)^{-\beta} \sim t^{-\beta \mu}, \tag{6.27}$$

which describes the dissipate behavior of a system. The values of $\beta \mu$ drops from 0.4 for PVAc to 0.1 for epoxy resins. As a result, a smaller effect of physical aging on the dynamic response of crosslinked polymers is expected. Comparing with linear polymers, we see that epoxy resins have smaller β and larger f_r and ε. The formation of crosslinks slows the molecular motion of chain molecules and the macroscopic relaxation time has to increase. This effect is also related to stronger cooperative interactions caused by a broader relaxation spectrum (smaller β) for networks, which contributes to more frozen-in free volume near the glass transition.

6.4   Yield Behavior

In solid-state deformation, the nonlinear viscoelastic effect is most clearly shown in the yield behavior. The type of stresses applied to a system has little effect on the linear viscoelastic response, in terms of the relaxation function (Φ) and time (τ), but becomes important as the stress level increases. At high stress levels, the contribution from the external work done on a lattice cell has to be included in the nonlinear viscoelastic analysis. Free volume (hole) has played a central role in the molecular interpretation of the local configuration rearrangements of molecular segments and the glassy-state relaxation. The change in free volume is also expected to have a strong influence on the yield stresses as on the glass transition.

Consider uniform stress tensor σ_{ij} acting on the polymer lattice. In order to overcome the local energy barrier during the yielding processes, the average work acting on each lattice site can in general be written as

$$\Delta w = -\sigma_{ij}\Omega_{ij}, \tag{6.28}$$

where Ω_{ij} is the activation volume tensor (see Section 6.6). It represents the volume of the polymer segments that has to move as a whole for the plastic yield to occur. The activation volume tensor plays a key role in nonlinear viscoelasticity. The relaxation time takes the form

$$\tau(T, \delta, \sigma_{ij}) = \tau_r a(T, \delta) \exp\left(-\frac{\sigma_{ij}\Omega_{ij}}{2\beta kT}\right). \tag{6.29}$$

It is a generalization of the relaxation time discussed in Section 5.4. The change in the physical mechanism of deformation from elasticity to viscoelasticity and to plasticity depends on the time scale in which the amorphous solid is measured and relaxed. The dependence of the nonlinear stress–strain relationships on the relaxation times is conceptualized in Figure 1.3, where the yield stresses are defined.

The yield occurs when the product of the relaxation time and the applied strain rate reaches a constant value [12,13]:

$$\dot{e}\tau \sim constant. \tag{6.30}$$

Using eqs. (6.28)–(6.30), and replacing σ_{ij} by $\sigma_y^{(ij)}$, we obtain the yield stress components

$$\sigma_y^{(ij)} = A_{ij} + K_{ij}[\log \dot{e} + \log a(T, \delta)], \tag{6.31}$$

where A_{ij} are constants and $K_{ij} \propto \beta kT / \Omega_{ij} \sim \Omega_{ij}^{-1}$. In addition to the well-known dependence of yield stress on temperature and strain rate, Eq. (6.31) provides a functional relationship between the plastic yield, the physical aging, and the type of stresses applied. Considering a glassy polymer under uniaxial tension, and substituting Eq. (5.46) into Eq. (6.31), we get

$$\Delta\sigma_y(T, t)/K = -\frac{\Delta\delta(T, t)}{2.303\beta f_r^2} \cong \mu \log\left(\frac{t}{t_o}\right). \tag{6.32}$$

For simplicity, the superscript and subscript for σ_y and K were dropped.

In Figure 6.8, we see the linear relationship between $\Delta\sigma_y$ and logarithmic aging time t in the glassy state $(T < T_r - 20\,\mathrm{K})$. No such simple relationship exists, however, in the glass transition region. The diminishing effect of physical aging is a result of vanishing δ as $T \to T_g$. The theoretical predictions in Figure 6.8 are consistent with experimental observations [14]. A strong resemblance exists between figures 6.3 and 6.8, as one may expect from Eq. (6.31). The relationship between $\Delta\sigma_y$ and the change in nonequilibrium glassy state $\Delta\delta$ at various annealing temperatures is shown in Figure 6.9. The points are experimental data, and the solid line represents Eq. (6.32) with $\beta f_r^2 = 0.804 \times 10^{-3}$ for polycarbonate.

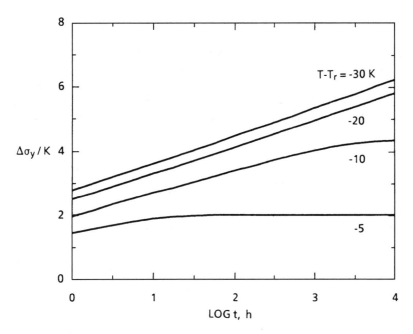

FIGURE 6.8. The dependence of the yield stress of PVAc on the physical aging time (t) and temperature in the vicinity of the glass transition is calculated from Eq. (6.32).

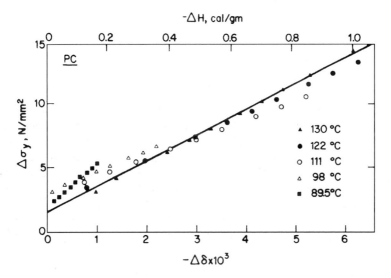

FIGURE 6.9. The change in the yield stress versus the change in the nonequilibrium state δ of polycarbonate. The line transition is calculated from Eq. (6.32) at $T = 130\,°C$, and the points are experimental data [15] at different annealing temperatures.

6.5   Stress-Induced Glass Transition

The effects of applied stress and stress rate on T_g will be analyzed in accordance with the nonequilibrium approach mentioned in Section 6.1. Let us return to the lattice model that consists of $N = n + x n_x$ sites, where $n = \sum_l n_l$ is the hole number and x is the number of monomer segments/polymer, and each site is associated with a unit cell of volume v. The entropy increase caused by the introduction of hole is equal to [see Eq. (4.42)]

$$\Delta S_m = -Nk \left[\sum_{l=1}^{L} f_l \ln f_l + (1/x)(1-f)\ln(1-f) \right]. \qquad (6.33)$$

The internal entropy of the hole has been neglected in the above equation, which is also assumed to be valid in nonequilibrium not too far from equilibrium. The time derivative of Eq. (6.33) is

$$\frac{d(\Delta S_m)}{dt} = -Nk \sum_{l=1}^{L} (1 + \ln f_l) \dot{f}_l \cong Nk \sum_{l=1}^{L} (1 + \ln \bar{f}_l) \frac{\delta_l}{\tau_l}, \quad \text{for} \quad 1/x \to 0,$$

$$(6.34a)$$

where Eq. (5.33) was used. The equilibrium free-volume fraction $\bar{f}_l$ shown in Eq. (6.34a) has the same form as Eq. (4.43), with ε_l being replaced by $\varepsilon_l + \Delta w$; i.e.,

$$\bar{f}_l = \frac{\bar{n}_l}{N} = c \exp\left(-\frac{\varepsilon_l + \Delta w}{kT}\right). \qquad (6.34b)$$

Thus,

$$\frac{d(\Delta S_m)}{dt} = -(N/T) \sum_{l=1}^{L} (\varepsilon_l + \Delta w) \frac{\delta_l}{\tau_l}. \qquad (6.35)$$

The entropy product-per-unit time inside the system is an inherent characteristic of structural relaxation in polymer glasses, in which part of the energy is irreversibly dissipated. The variation in the entropy may be analyzed as the sum of two contributions. One contribution is from the entropy produced inside the system, and the other is supplied to the system by the applied stress field. The latter results in the deformational contribution to the entropy change during the glass formation:

$$\Delta S_{def} = -\partial(\Delta F)/\partial T |_{T_r} = \Delta \alpha \sigma V/3 = \frac{(\varepsilon + \Delta w)\Delta \kappa}{3T_r} \sigma N, \qquad (6.36)$$

where F is the strain energy [16] and $\sigma = \sigma_{11} + \sigma_{22} + \sigma_{33}$ is the stress invarient. Eq. (5.8) was used in Eq. (6.36). The entropy product-per-unit time can therefore be written as

$$\frac{d(\Delta S_m)}{dt} = -(N/T) \sum_{l=1}^{L} (\varepsilon_l + \Delta w)(\dot{f}_l + \Delta \kappa_l \cdot \dot{\sigma}/3). \qquad (6.37)$$

Comparing eqs. (6.35) and (6.37) yields

$$d\delta_l/dt = -\delta_l/\tau_l - \Delta\alpha_l \cdot q - \Delta\kappa_l \cdot \dot{\sigma}/3, \qquad l = 1, \dots, L. \qquad (6.38)$$

It is a generalization of Eq. (5.34) in the vicinity of T_g.

The nonequilibrium criterion described in Section 6.1 is now going to be extended for the determination of the stress-induced glass transition. Following Eq. (6.2) and Eq. (6.38), we have

$$d\delta/dt = -\delta/\langle\tau\rangle - \Delta\alpha \cdot q - \Delta\kappa \cdot \dot{\sigma}/3. \qquad (6.39)$$

In terms of the fictive temperature, this equation can be approximated by

$$T_f = T + \left(|q| - \frac{\Delta\kappa}{\Delta\alpha}\frac{\dot{\sigma}}{3}\right)\langle\tau\rangle \qquad (6.40)$$

for a system cooling from a equilibrium liquid temperature. The glass transition temperature is then determined by the freezing-in condition: $dT_f/dT = 0$. When all higher order terms, $O(T_g^{-1})$, are neglected, we obtain

$$\frac{d\langle\tau\rangle}{dT} = -\left(|q| - \frac{\Delta\kappa}{\Delta\alpha}\frac{\dot{\sigma}}{3}\right)^{-1}, \qquad \text{at } T = T_g. \qquad (6.41)$$

This expression is a generalization of Eq. (6.5) — the nonequilibrium criterion for T_g of amorphous polymers vitrified under cooling and applied stress. The solution of Eq. (6.29) and Eq. (6.41) is

$$T_g = T_r + \frac{\beta f_r^2}{\Delta\alpha} \ln\left[\frac{\tau_r \Delta\alpha}{\beta f_r^2}\left(|q| - \frac{\Delta\kappa}{\Delta\alpha}\frac{\dot{\sigma}}{3}\right)\right] - \frac{1}{\Delta\alpha}\delta|_{T=T_g} + \frac{\beta f_r^2 \Delta w}{2kT_g\Delta\alpha}. \qquad (6.42)$$

When a system is under a uniform stress field with negligible stress rate, the effect of applied stress on T_g can be evaluated by

$$T_g - T_{go} = -\frac{1}{\Delta\alpha}(\delta - \delta_o)|_{T=T_g} + \frac{\beta f_r^2}{\Delta\alpha}\ln\left(1 + \frac{\Delta w}{\varepsilon}\right) - \frac{\beta f_r^2 \Delta w}{2kT_{go}\Delta\alpha}, \qquad (6.43)$$

simplified from Eq. (6.42). Here, the subscript "o" refers to the unstressed state, and Δw is positive for compression and negative for tension.

In general, $|\Delta w| \ll \varepsilon$ and the last two terms on the right-hand side of Eq. (6.43) are found to be numerically negligible in comparison with the first term. When $t \gg \tau(T = T_g)$, the approximate solution of Eq. (6.39) at T_g is $\delta - \delta_o \approx \bar{f} - \bar{f}_o$. By using eqs. (6.34b) and (6.43), the effect of stresses on T_g is obtained

$$T_g = T_{go} + \frac{f_r}{\Delta\alpha}\left[1 - \exp\left(-\frac{\Delta w}{kT_{go}}\right)\right]. \qquad (6.44)$$

This equation is applicable to any stress conditions, which may be dilatational, shear, or their combinations. The functional behavior of T_g and Δw is plotted in Figure 6.10. The level of extension is limited by the ultimate tensile strength of polymers, where $-\Delta w$ is usually less than kT_{go}. Clearly, compression causes

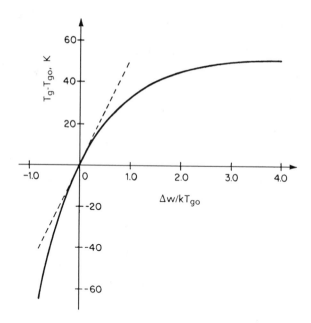

FIGURE 6.10. The change of the glass transition temperature as a function of the normalized work Δw due to the external stress field. $\Delta w > 0$ for compression and $\Delta w < 0$ for tension.

an increase in T_g, and tension reduces T_g. In contrast to the prevalent thinking, Figure 6.10 shows that T_g does not continue to increase at all pressures ($\Delta w = pv$) but levels off to a universal asymptote at very high pressure.

When $\Delta w / kT_{go} \ll 1$, the right-hand side of Eq. (6.44) can be expanded into a series. Consider poly(vinyl chloride) (PVC), the measured pressure coefficient near the glass transition $T_{go} = 360K$ is $dT_g/dp = vT_r/\varepsilon = 0.016K/\text{bar}$ [17]. When a system is under uniform tensile stress (σ_{11}), we have $\Delta w = -\sigma_{11}\Omega_{11}(+)$ and $dT_g/d\sigma_{11} = -\Omega_{11}(+)T_r/\varepsilon$ [see Eq. (6.53)]. The theoretical calculation and the uniaxial-creep measurement of the stress-induced T_g for PVC are compared in Figure 6.11, where the dotted line is the linear approximation. From these two slopes, the ratio of the lattice volume to the tensile activation volume is determined:

$$\frac{v}{\Omega_{11}(+)} = 0.172. \tag{6.45}$$

This expression has been experimentally verified for many amorphous polymers, including polycarbonate, PVAc, and polystyrene.

Finally, let us look at a system subjected to dynamic tests while holding the static load and temperature unchanged. From Eq. (6.42), the glass transition temperatures T_{g1} and T_{g2} are related to the stress rates $\dot{\sigma}_1$ and $\dot{\sigma}_2$ by

$$T_{g2} - T_{g1} = \frac{\beta f_r^2}{\Delta \alpha} \ln\left(\frac{\dot{\sigma}_2}{\dot{\sigma}_1}\right). \tag{6.46}$$

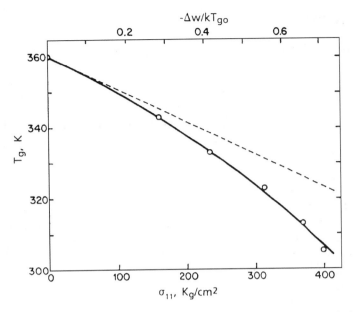

FIGURE 6.11. A comparison of the calculated (solid curves) and measured (circles [18]) glass transition temperature versus the tensile stress of poly(vinyl chloride).

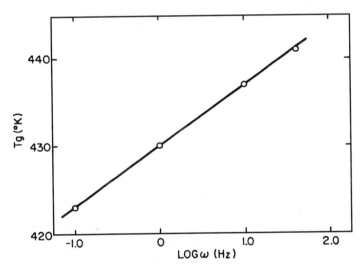

FIGURE 6.12. A comparison of the calculated (solid curves) and measured (circles [19]) glass transition temperature versus log (frequency) for carbon-epoxy resin.

Applying oscillating stresses to the sample and holding the amplitude of dynamic loading constant, we have $\dot{\sigma}_2/\dot{\sigma}_1 = \omega_2/\omega_1$. Replacing the ratio of stress rate in Eq. (6.46) by the ratio of frequencies provides a good description of the dynamic viscoelastic measurements of carbon-epoxy resin in Figure 6.12, with $\beta f_r^2/\Delta\alpha = 2.94$ K, which is a reasonable value [see Eq. (5.67)].

6.6   Activation Volume Tensor

The new tensorial extensive quantity introduced in Eq. (6.28) is defined by

$$\Omega_{ij} = V\left(e_{ij}^{liquid} - e_{ij}^{glass}\right)/n \equiv v\Delta e_{ij}/f, \tag{6.47}$$

so that the average work acting on the hole cell during the yielding is Δw. It represents the volume of the polymer segments that has to move as a whole for plastic yield to occur. For $i = j$, the excess strain invariant $\Delta e_{ii} = f$ and Eq. (6.47) reduces to

$$\Omega_{ii} = v. \tag{6.48}$$

Here, Ω_{ii} is called the pressure activation volume. This equation provides a consistent check that the pressure activation volume is equal to the volume of a single lattice cell. It is often convenient to split a second-order stress tensor (σ_{ij}) into its scalar (σ) and deviatoric ($'\sigma_{ij}$) parts. Yield occurs when the magnitude of shear stress reaches a critical value. Thus, we write the absolute value of the stress deviatoric components

$$'\sigma_{ij} = \left|\sigma_{ij} - \frac{\sigma}{3}\delta_{ij}\right|, \tag{6.49}$$

where σ is the stress invariant, and δ_{ij} is the Kronecker delta. Eq. (6.49) leads to

$$\sigma_{ij} = \frac{\sigma}{3}\delta_{ij} + '\sigma_{ij} = \frac{\sigma}{3}\delta_{ij} + \left|\sigma_{ij} - \frac{\sigma}{3}\delta_{ij}\right|. \tag{6.50}$$

Similarly, the activation volume tensor can be slit into two parts

$$\Omega_{ij} = \frac{v}{3}\delta_{ij} + '\Omega_{ij}. \tag{6.51}$$

Using eqs. (6.28), (6.50), and (6.51), the work contributed by external stresses acting on the lattice is determined by

$$-\Delta w = \frac{\sigma v}{3}\delta_{ij} + '\sigma_{ij}'\Omega_{ij} = \frac{\sigma v}{3}\delta_{ij} + \Omega_{12}\left|\sigma_{ij} - \frac{\sigma}{3}\delta_{ij}\right|, \tag{6.52}$$

where the usual summation convention for repeated suffixes in $'\sigma_{ij}'\Omega_{ij}$ over values 1, 2, and 3 is followed.

For isotropic glasses, the activation volume tensor has two independent components. These components are the bulk activation volume, which is equal to v, and the shear activation volume Ω_{12}. In the case of uniaxial tension (σ_{11}), Eq. (6.52) becomes

$$-\Delta w = \sigma_{11}\frac{2\Omega_{12} + v}{3} \equiv \sigma_{11}\Omega_{11}(+). \tag{6.53}$$

In uniaxial compression ($-\sigma_{11}$), one gets

$$-\Delta w = \sigma_{11}\frac{2\Omega_{12} - v}{3} \equiv \sigma_{11}\Omega_{11}(-). \tag{6.54}$$

Therefore,

$$\Omega_{11}(\pm) = \frac{2\Omega_{12}}{3}\left(1 \pm \frac{v}{2\Omega_{12}}\right). \qquad (6.55)$$

We have already determined the ratio of the lattice volume to the tensile activation volume by Eq. (6.45). Combining eqs. (6.45) and (6.55) gives

$$\Omega_{11}(+) : \Omega_{11}(-) : \Omega_{12} = 1 : 0.885 : 1.42. \qquad (6.56)$$

It has been observed experimentally that the yield stress decreases with an increase in temperature in accordance with $\sigma_y^{(ij)} = C_{ij}(T_s - T)$, where T_s is the softening temperature and C_{ij} is insensitive to strain rate. From Eq. (6.45), the negative slope of yield stress components versus temperature is

$$-\frac{d\sigma_y^{(ij)}}{dT} = C_{ij} = \frac{\varepsilon}{\Omega_{ij} T_r}. \qquad (6.57)$$

Using the $\sigma_y - T$ data of polystyrene in compression at a constant strain rate, we obtain $C_{11}(-) = \varepsilon/T_r\Omega_{11}(-) = -6.8$ kg/cm^2 K from Figure 6.13. We have

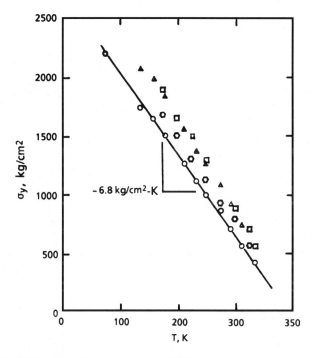

FIGURE 6.13. The relationship between the yield stress and temperature of polystyrene under uniaxial compression. The line represents Eq. (6.57) and the points are experimental data [20].

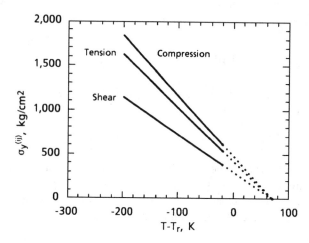

FIGURE 6.14. The predicted temperature dependence of the yield stresses of polystyrene under uniaxial tension, compression, and shear.

determined $\varepsilon = 3.58$ kcal/mol and $T_r = 370.5 K$ for polystyrene (PS) [21]. Hence,

$$[\Omega_{11}(+) : \Omega_{11}(-) : \Omega_{12} : v] = [111.7 : 98.8 : 158.6 : 19.2]\mathring{A}^3 \qquad (6.58)$$

Except in the case of hydrostatic compression, Eq. (6.58) reveals that the volume of molecular segments that has to move as a whole at yielding is much larger than the volume of a single lattice site. The motion of the polymer segments is no longer local, which is in contrast to the situation in linear viscoelastic responses. This result reinforces our earlier assumption that the yielding is a cooperative phenomenon. By using Figure 6.13, eqs. (6.57) and (6.58), the effect of stress field on the yield stresses as a function of temperature is predicted in Figure 6.14. Shear is a signature of the onset of plastic yield. For $T \geq T_s$, the mechanical properties are no longer determined by the configurational rearrangements of molecular segments. Large-scale motion of polymer chains becomes important, and the polymer flow, instead of deformation, plays a major role.

As we have seen in Eq. (6.31), the dependence of yield stress on strain rate under uniaxial compression for PS at the room temperature is determined by the parameter: $K_{11}(-) = 90$ kg/cm^2 [22]. By using eqs. (6.31) and (6.56), the relationships between the yield stress and strain rate for three different stresses applied to the system are shown in Figure 6.15.

6.7   Nonlinear Stress–Strain Relationships

This topic is one of the most important mechanical properties of solid polymers and has been a subject of intense investigation over years. The distinct features of glassy polymers are that their physical properties vary more strongly with time and temperature than those of metal and ceramics. As we have already mentioned,

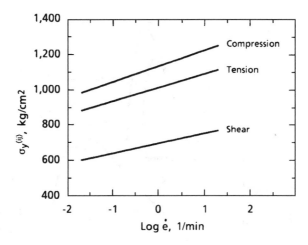

FIGURE 6.15. The predicted strain-rate dependence of the yield stresses of polystyrene under uniaxial compression tension, and shear at 23 °C.

the long-time behavior and thermal history are among the major concerns. The important influence of the glassy-state relaxation forms the basis of analyzing the deformation kinetics in the glassy state depicted in Figure 1.3. The functional relationships between the structural relaxation and nonlinear deformation established so far in this chapter will enable us to make some quantitative prediction of the nonlinear stress–strain relationships as a function of physical aging, strain rate, temperature, and external stress field.

The stress-dependent relaxation modulus of glassy polymers can be written in the form of Eq. (6.12) as

$$C_{ijkl}(t, \sigma_{ij}) = C_{ijkl}^{(0)} \exp\{-[t/\tau(\sigma_{ij})]^{\beta}\}, \qquad 0 < \beta \le 1, \qquad (6.59)$$

where the stress-dependent τ is given by Eq. (6.29), and $C_{ijkl}^{(0)}$ is the elastic moduli. The above equation will be treated as the nonlinear relaxation function for high stresses. For nonlinear deformation, the dependence of the relaxation moduli on the stress components in Eq. (6.59) results in the continuous change of C_{ijkl}, which defines the slope of the stress–strain curve with other parameters held constant. Thus, putting $t = e_{kl}/\dot{e}_{kl}$ and integrating Eq. (6.59), we get a nonlinear integral equation:

$$\sigma_{ij}(e_{kl}) = C_{ijkl}^{(0)} \int_{o}^{e_{kl}} \exp\left\{ -\left[\frac{e_{kl}' \exp(2.303\sigma_{kl}(e_{kl}')/K_{kl})}{\dot{e}_{kl}\tau_o} \right]^{\beta} \right\} de_{kl}', \qquad (6.60)$$

where $\tau_o \equiv \tau(\sigma_{ij} = 0)$, the summation convention is not used, and the strain rate is kept constant. Eq. (6.60) is particularly useful in interpreting the data generated from the Instron measurements. For isotropic glasses, only two independent components of $C_{ijkl}^{(0)}$ exist. The shear modulus (G) is related to Young's modulus (E) and Poisson's ratio (ν) by $G = E/2(1 + \nu)$.

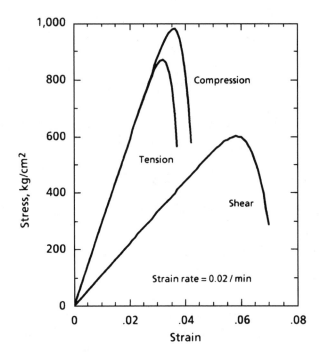

FIGURE 6.16. The calculated nonlinear stress–strain behavior of polystyrene under uni-axial tension, compression, and shear at 23 °C.

In addition to K_{ij}, mentioned in the last section, $E = 29.6 \times 10^3$ kg/cm^2, $\nu = 1/3$, $\beta = 0.48$, and $\tau_0 = 2 \times 10^{14}$ min at 23 °C for a well-aged PS are adopted in seeking the numerical solutions of Eq. (6.60). The calculated effect of stress field on the nonlinear stress–strain behavior is shown in Figure 6.16. The theoretical tensile and compressive stress–strain curves compare well with the reported data [23]. We see that σ_y(compression) $> \sigma_y$(tension) $> \sigma_y$(shear), and the onset of yielding begins with shear. The effect of strain rate is calculated in Figure 6.17. The maxima in Figure 6.17 compare exactly with the yield stresses shown in Figure 6.15, which are obtained directly from the experimentally verified Eq. (6.31).

In the study of physical aging, let us focus the discussion on the uniaxial com-pression. The subscript "11" for stress or strain is dropped. We have already seen that the relaxation time, Eq. (6.29), of a quenched and annealed glass depends strongly on the absolute time that changes not only with the annealing time (t), but also with the loading time. Both t and the loading time are usually coupled. The load time ($e/\dot{e}$) in the Instron measurement, however, is always many orders of magnitude smaller than the aging time and can be neglected in the calculation of the physical-aging effect.

The compressibility of the polymer lattice, expressed by xn_x, does not contribute to the physical aging, but the contribution from the hole in the vicinity of the glass transition can be approximated as κ(hole) $\cong \mathrm{v}(\bar{f} + \delta)/kT$ (see Section 5.1). The

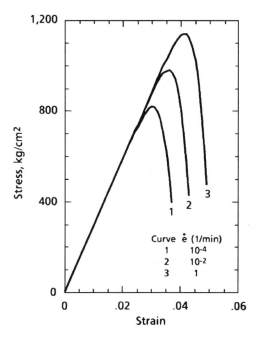

FIGURE 6.17. Dependence of the compressive yield stress of polystyrene on strain rate. The sample has been well aged ($t_a = 10^4 h$ at $T = T_r - 20$ K).

Young modulus is inverse proportional to the compressibility. Consider that a system is cooled from liquid to glass and then annealed isothermally. By using Eq. (5.44), the change in Young's modulus is obtained

$$\Delta E_0 \equiv E_0(t_2) - E_0(t_1) = 3\beta(1 - 2v)\frac{kT}{v}\Delta \ln a(T, \delta) \qquad (6.61)$$

for polymers aged from $t = t_1$ to t_2 at the temperature T in the glassy state. The effects of the aging time and temperature on the change in Young's modulus of a quenched and annealed PS are calculated in Figure 6.18. The shape of the curves depends on how long it takes for the nonequilibrium state (δ) to approach its equilibrium at different annealing temperatures (also see Figures 6.3 and 6.8). The input hole parameters for PS with $T_r = 370.5$ K are summarized:

$$\varepsilon = 3.58 \text{ kcal/mol}, \quad f_r = 0.032, \quad \beta = 0.48, \quad \tau_r = 30 \text{ min}. \qquad (6.62)$$

They are used in the calculation of the volume relaxation required in Eq. (6.61) and the Young modulus shown in Figure 6.18. By using Eq. (6.61) for $E_0(t_a)$ and Eq. (5.46) for τ_o in Eq. (6.60), the effect of physical aging on the nonlinear compressive stress–strain behavior is calculated in Figure 6.19. As the aging time increases, both Young's modulus and the yield stress increase but the aged polymer becomes more brittle.

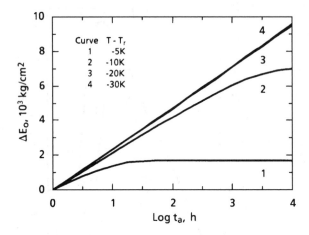

FIGURE 6.18. Effect of annealing time (t) on the change in Young's modulus of a polystyrene glass. Curves represent the calculations at different annealing temperatures.

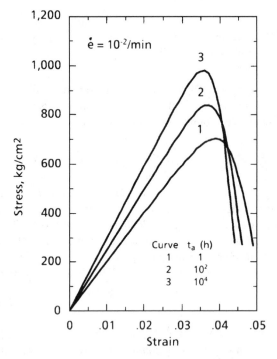

FIGURE 6.19. Effect of physical aging on the nonlinear stress–strain curves of polystyrene under uniaxial compression. The annealing temperature is $T = T_r - 20$ K.

References

1. J. H. Gibbs and E. A. DiMarzio, J. Chem. Phys. **28**, 373 (1958).
2. P. J. Flory, Proc. R. Soc. London **A234**, 60 (1956).
3. T. S. Chow, Macromolecules **22**, 698 (1989).
4. A. Q. Tool, J. Am. Ceram. Soc. **29**, 240 (1946).
5. R. O. Davis and G. O. Jones, Proc. R. Soc., London **A47**, 26 (1953).
6. T. S. Chow, Polym. Eng. Sci. **24**, 1079 (1984).
7. L. C. E. Struik, *Physical Aging in Amorphous Polymers and Other Materials* (Elsevier, Amsterdam, 1978).
8. W. G. Knauss and V. H. Kenner, J. Appl. Phys. **51**, 5131 (1980).
9. J. C. Halpin, in *Composite Materials Workshop*, edited by S. W. Tsai, J. C. Halpin, N. J. Pagano (Technomic, Stanford, CT, 1968).
10. D. H. Kaeble, J. Appl. Poly. Sci. **9**, 1213 (1965).
11. T. S. Chow, J. Polym. Sci. **B25**, 137 (1987).
12. I. M. Ward, *Mechanical Properties of Solid Polymers*, 2nd ed. (Wiley, New York, 1983).
13. H. Eyring, J. Chem. Phys. **4**, 283 (1936).
14. C. G'Sell and G. B. McKenna, Polymer, **33**, 2103 (1992).
15. H. J. Ott, Colloid Polym. Sci. **258**, 995 (1980).
16. L. D. Landau and E. M. Lifshitz, *Theory of Elasticity* (Pergamon, Oxford, 1959).
17. M. C. Shen and A. Eisenberg, Prog. Solid State Chem. **3**, 407 (1966).
18. R. D. Andrews and Y. Kazama, J. Appl. Phys. **38**, 4118 (1967).
19. S. S. Sternstein, Am. Chem. Soc., Polym. Prepr. **22**, 237 (1981).
20. J. P. Cavrot, J. Haussy, J. M. Lefebvre, and B. Escaig, Mater. Sci. Eng. **26**, 95 (1978).
21. T. S. Chow, J. Rheol. **36**, 1707 (1992).
22. T. S. Chow, Polymer **34**, 541 (1993).
23. R. N. Howard, B. M. Murphy, and E. F. T. White, J. Polym. Sci. A-2, **9**, 801 (1971).

7

Polymer Composites

Composite materials may be considered as materials made of two or more components and consisting of two or more phases in the solid state. The importance of polymer composite [1] originates largely because such low-density materials can have unusual high elastic constants and tensile strength. For the most part, the tensile properties have been adequately dealt with by using the theory of elasticity. With new demands for materials to survive in a severe environment of high temperature, compression, and shear rate, deformation from elasticity to viscoelasticity and to plasticity becomes important as the loading and environmental conditions vary.

One purpose of this chapter is to provide a unified understanding and simple approach to the effective elastic moduli, thermal expansion coefficients, stress concentrations, yield stresses, and nonlinear stress–strain behaviors of particulate composites. In the first part of this chapter, an equilibrium composite theory will be presented on the basis of an effective medium theory that takes into the account of particle–particle interactions for systems consisting of nonspherical particles dispersed randomly in polymers. Several effective properties of composites are expressed by the elastic moduli of particle and matrix, and the aspect ratio, shape, and orientation of particles. The glassy-state relaxation (see chapters 5 and 6) will then be incorporated into the composite theory to describe the nonequilibrium mechanical properties of polymer composites.

Beyond the above-mentioned particulate composites, we shall discuss in the last three sections some of the unusual phenomena observed for molecular composites below the glass transition temperature. These phenomena include the stress anomaly in compatible polymer blends and the order–disorder transition in nanocomposites. This discussion helps us to see the limitation of the traditional composite theories and the important role of molecular parameters.

7.1   Anisotropic Elasticity

The effective elastic constants of a composite containing either spherical particles or reinforced fibers, dispersed randomly in a matrix, has been a subject of intensive investigations because of its importance. When a system is filled with nonspherical-oriented particles, the composite becomes anisotropic in nature. Many equations have been developed [2,3]; however, we would like to present a simple but useful approach that has the capability of covering a broad range of expressions from disk-like particles to spheres to continuous long fibers by analyzing aligned ellipsoidal particles rigorously. The anisotropic effect is characterized by the ratio of major to minor axes ($\rho = c/a$). The many-body problems of particle-to-particle interactions at finite concentrations are handled using an effective medium theory (see Section 4.2). In the succeeding sections, it will become clear that this approach is not limited to the problem of anisotropic elastic moduli but is extensive in the study of other physical properties.

Analyses of elastic constants of composites usually require the determination of the elastic field around fillers, which involves a tedious boundary value problem for nonspherical particle and becomes even more impracticable when the particle–particle interactions are included. At dilute concentration, the elegant work of Eshelby [4,5] is well known for its simplicity in analyzing the elastic constants without getting into the unnecessary mathematical complication.

Let us consider uniform strain tensor e_{ij}^A acting on a two-phase system containing a spheroid. The strain tensor in the particle is

$$e_{ij}(f) = e_{ij}^C + e_{ij}^A, \tag{7.1}$$

where e_{ij}^C is the constrained strain tensor in the particle and f refers to the filler (particle). By using Hooke's law, the corresponding stress tensor is [6]

$$\sigma_{pq}(f) = C_{pqij}^{(f)}\left(e_{ij}^C + e_{ij}^A\right), \tag{7.2}$$

where $C_{pqij}^{(f)}$ are the elastic constants of a filler. The usual summation convention is followed for repeated suffixes over values 1, 2, and 3. Eshelby showed that the inhomogeneity problem can be reduced to that of transformation in an equivalent particle whose elastic moduli are equal to those of the matrix, $C_{pqij}^{(m)}$. That is,

$$\sigma_{pq}(f) = C_{pqij}^{(m)}\left(e_{ij}^C + e_{ij}^A - e_{ij}^T\right), \tag{7.3}$$

where e_{ij}^T is the transformation strain tensor in the equivalent particle and m refers to the matrix (polymer). The constrained strain tensor is related to the transformation strain tensor by [4]

$$e_{ij}^c = S_{ijkl}e_{kl}^T. \tag{7.4}$$

Here, the Eshlby tensor S_{ijkl} is a function of the aspect ratio of a spheroid and Poisson's ratio v_m of the matrix (see Appendix 7A). The particle-shape tensor that appeared in the theory of liquid [see Eq. (4.15)] looks like Eshlby's tensor in Eq. (7.4). The difference is that the particle-shape tensor is independent of material parameter. Equating the right-hand sides of eqs. (7.2) and (7.3), six equations are derived

$$C_{pqij}^{(f)}\left(e_{ij}^C + e_{ij}^A\right) = C_{pqij}^{(m)}\left(e_{ij}^C + e_{ij}^A - e_{ij}^T\right), \tag{7.5}$$

for the six unknowns e_{ij}^T. e_{ij}^C can be eliminated from Eq. (7.5) by using Eq. (7.4). For simplicity, both the particle and matrix are assumed to be isotropic. That is,

$$C_{pqij} = \left(K - \tfrac{2}{3}G\right)\delta_{pq}\delta_{ij} + G(\delta_{pi}\delta_{qj} + \delta_{pj}\delta_{qi}), \tag{7.6}$$

where K and G are the bulk and shear moduli, respectively, and δ_{ij} is the Kronecker delta.

The anisotropy is a result of a nonspherical filler at all concentrations. When fillers are considered as identical spheroids

$$\frac{x_1^2 + x_2^2}{a^2} + \frac{x_3^2}{b^2} = 1, \tag{7.7}$$

with corresponding axes aligned, the heterogeneous medium becomes transversely isotropic about the x_3 direction and the composite as a whole can then be described by five independent moduli. They are one bulk modulus (K), two shear moduli ($G_{23} = G_{13}$, and G_{12}), and two Young's moduli (E_{33} and $E_{11} = E_{22}$), as shown in Figure 7.1. The constitutive relation between the strain and stress tensors of the

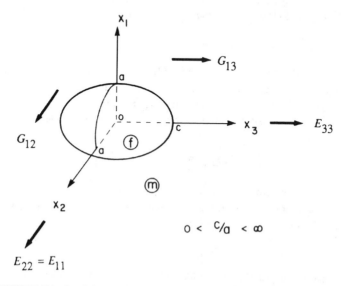

FIGURE 7.1. Particle geometry, orientation, notation and elastic constants.

particulate composite can be expressed by the following equations:

$$e_{11} = \frac{1}{E_{11}}\sigma_{11} - \frac{\nu_1}{E_{11}}\sigma_{22} - \frac{\nu_3}{E_{33}}\sigma_{33}, \tag{7.8a}$$

$$e_{22} = -\frac{\nu_1}{E_{11}}\sigma_{11} + \frac{1}{E_{11}}\sigma_{22} - \frac{\nu_3}{E_{33}}\sigma_{33}, \tag{7.8b}$$

$$e_{33} = -\frac{\nu_3}{E_{33}}\sigma_{11} - \frac{\nu_3}{E_{33}}\sigma_{22} + \frac{1}{E_{33}}\sigma_{33}, \tag{7.8c}$$

and

$$e_{ij} = \sigma_{ij}/2G_{ij}, \qquad ij = 23, 13, 12. \tag{7.8d}$$

The Poisson ratios are

$$\nu_1 = E_{11}/2G_{12} - 1 \tag{7.9a}$$

and

$$\nu_3 = \tfrac{1}{2}(1 - E_{33}/3K). \tag{7.9b}$$

Because K and G are always positive, the two Poisson ratios have to be with the physical limits of

$$-1 \le \nu_i \le \tfrac{1}{2}, \qquad i = 1, 3. \tag{7.10}$$

7.2 Elastic Constants

At the finite volume fraction of filler ϕ, the interactions between particles have to be included in the analysis to derive the five independent elastic constants of composites containing aligned spheroids with the aspect ratio $\rho = c/a$. The particles are assumed to be uniform in size and firmly bonded to the matrix. Besides ϕ and ρ, the effective elastic moduli are going to be expressed by the isotropic moduli for the filler (G_f, K_f) and for the matrix (G_m, K_m). The anisotropy is a result of $\rho \ne 1$.

When an unperturbed uniform shear strain e_{ij}^A is applied to the system, the local strain in the matrix or particles is

$$e_{ij}(\vec{r}) = e_{ij}^A + e_{ij}^C(\vec{r}). \tag{7.11}$$

When the spatial distribution of aligned particles is random and homogeneous, the composite as a whole has to be macroscopically homogeneous; i.e.,

$$\frac{1}{V}\int_V e_{ij}(\vec{r})\, dV - e_{ij}^A = 0. \tag{7.12}$$

Substituting Eq. (7.11) into Eq. (7.12) gives

$$\frac{1}{V}\left[\int_{V_m} e_{ij}^C(\vec{r})\, dV + N\int_{V_f} e_{ij}^C(\vec{r})\, dV\right] = 0, \tag{7.13}$$

where N is the number of particles in the total volume $V = V_m + NV_f$. By defining the volume averages,

$$\langle e_{ij}^C \rangle_\ell = \frac{1}{V_\ell} \int_{V_\ell} e_{ij}^C(\vec{r})\,dV, \qquad \ell = f \text{ or } m, \tag{7.14}$$

Eq. (7.13) becomes

$$\phi \langle e_{ij}^C \rangle_f + (1 - \phi)\langle e_{ij}^C \rangle_m = 0. \tag{7.15}$$

Here, $\phi = NV_f/V$ is the volume fraction of the particle. The average constrained strain $\langle e_{ij}^C \rangle_m$ set up in the matrix is caused by the interactions of particles when their volume fraction is high. As $\phi \to 0$, $\langle e_{ij}^C \rangle_m$ should approach zero. Therefore, in the extension of Eq. (7.4) beyond the dilute limit, it is reasonable to assume that [7]

$$\langle e_{ij}^C \rangle_f = S_{ijkl}\langle e_{kl}^T \rangle_f + \langle e_{ij}^C \rangle_m. \tag{7.16}$$

Eliminating $\langle e_{ij}^C \rangle_m$ between eqs. (7.15) and (7.16), we obtain the relation between the average constrained and transformation strains:

$$\langle e_{ij}^c \rangle_f = (1 - \phi)S_{ijkl}\langle e_{kl}^T \rangle_f. \tag{7.17}$$

By generalizing Eq. (7.5), the volume average of stresses in a particle must satisfy the condition:

$$C_{pqij}^{(f)}\left(\langle e_{ij}^C \rangle_f + e_{ij}^A\right) = C_{pqij}^{(m)}\left(\langle e_{ij}^C \rangle_f + e_{ij}^A - \langle e_{ij}^T \rangle_f\right). \tag{7.18}$$

The summation over the repeated subscripts applies to the above equation. Eqs. (7.17) and (7.18) can be combined by eliminating the constrained strains. Using Eq. (7.6) for isotropic filler and matrix, we obtain five independent equations for $\langle e_{ij}^T \rangle_f$. They are

$$(1 - \phi)[(\lambda_f - \lambda_m)S_{kkpq}\delta_{ij} + 2(G_f - G_m)S_{ijpq}]\langle e_{pq}^T \rangle_f + \lambda_m \langle e^T \rangle_f \delta_{ij} + 2G_m\langle e_{ij}^T \rangle_f$$
$$= (\lambda_m - \lambda_f)e^A\delta_{ij} + 2(G_m - G_f)e_{ij}^A, \tag{7.19}$$

where $\lambda = K - 2G/3$ is the Lame constant and $e = e_{11} + e_{22} + e_{33}$ is the dilatation.

7.2.A   Shear Moduli

The effective shear moduli of a particulate composite are related to $\langle e_{ij}^T \rangle_f$ by [see Eq. (4.25)]

$$G_{ij} = G_m\left(1 - \frac{\langle e_{ij}^T \rangle_f}{e_{ij}^A}\phi\right), \qquad ij = 12, 13. \tag{7.20}$$

The average transform strains can be determined from Eq. (7.19) by the uniform applied strains e_{ij}^A. Because S_{ijkl} is not coupled in the case of shears, a straightforward

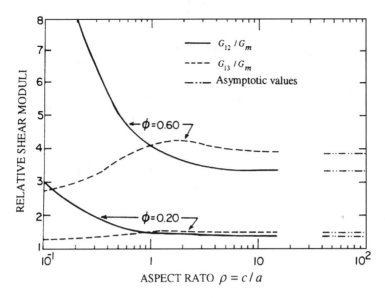

FIGURE 7.2. Dependence of the shear modulus on the particle shape, orientation and volume fraction (ϕ).

calculation leads to

$$\frac{G_{ij}}{G_m} = 1 + \frac{(G_f/G_m - 1)\phi}{1 + 2(G_f/G_m - 1)(1 - \phi)S_{ijij}}. \tag{7.21}$$

The explicit expressions of S_{1212} and S_{1313} are listed in Appendix 7A. When the aspect ratio $\rho = 1$, $G_{12} = G_{13} = G$ and Eq. (7.21) reduces exactly to the Kerner equation [8],

$$\frac{G}{G_m} = 1 + \frac{(G_f/G_m - 1)\phi}{1 + \chi(G_f/G_m - 1)(1 - \phi)}, \tag{7.22}$$

with $\chi = 2S_{1212} = 2S_{1313} = \frac{2}{15}(4 - 5v_m)/(1 - v_m)$. The experimentally verified Kerner equation was derived by a completely different and elaborate method that considers the composite as a spherical particle encased in a spherical shell of a matrix, which in turn is encased in an unbounded medium possessing the as-yet-unknown moduli of the composite. Having Eq. (7.22) as a special case of Eq. (7.21) demonstrates the success and simplicity of this effective medium theory in its dealing with the many-body problems. When $\rho \to \infty$, we have $2S_{1212} = (3 - 4v_m)/[4(1 - v_m)]$ and $2S_{1313} = 1/2$, which result in the asymptotic G_{ij} of a reinforced composite (see Figure 7.2).

7.2.B   Bulk Modulus

A similar calculation can be extended to the effective bulk modulus by considering a uniformed applied field

$$e_{11}^A = e_{22}^A = e_{33}^A = e^A/3. \tag{7.23}$$

The result is

$$\frac{K - K_m}{K_m \phi} = -\frac{\langle e^T \rangle_f}{e^A} = \frac{(K_f/K_m - 1)(C_1 + 2C_3)}{2B_1C_3 + C_1B_3}, \tag{7.24}$$

where

$$B_i = 1 + (K_f/K_m - 1)(1 - \phi)\vartheta_i, \qquad i = 1, 3$$

and

$$C_i = 1 + (G_f/G_m - 1)(1 - \phi)\chi_i, \qquad i = 1, 3.$$

Both ϑ_i and χ_i are functions of S_{ijkl}, which in turn depend on the aspect ratio ρ and Poisson's ratio ν_m of the matrix (see Appendix 7A). For spherical particles, $B_1 = B_3$, $C_1 = C_3$, and Eq. (7.24) assumes the corresponding form of the Kerner equation again.

7.2.C Young's Moduli

Consider a uniform tensile strain e_{33}^A with $e_{11}^A = e_{22}^A = 0$. This strain results in the longitudinal Young's modulus

$$\frac{E_{33} - E_m}{E_m \phi} = -\frac{\langle e_{33}^T \rangle_f}{e_{33}^A} = \frac{(K_f/K_m - 1)C_1 + 2(G_f/G_m - 1)B_1}{2B_1C_3 + C_1B_3}. \tag{7.25}$$

Applying a uniform strain e_{11}^A with $e_{22}^A = e_{33}^A = 0$, we obtain the transverse Young's modulus

$$\frac{E_{11} - E_m}{E_m \phi} = -\frac{\langle e_{11}^T \rangle_f}{e_{11}^A} = \frac{(K_f/K_m - 1)C_3 + 2(G_f/G_m - 1)(C_3\xi + B_3\zeta)}{2B_1C_3 + C_1B_3}, \tag{7.26}$$

where

$$\xi = \frac{B_1}{1 + 2(G_f/G_m - 1)(1 - \phi)S_{1212}}$$

and

$$\zeta = \frac{1 + (G_f/G_m - 1)(1 - \phi)(S_{1111} - S_{3311})}{1 + 2(G_f/G_m - 1)(1 - \phi)S_{1212}}.$$

To illustrate the anisotropic behavior of the five effective moduli varying from disk-like particles ($\rho \to 0$) to continuous long fibers ($\rho \to \infty$), we consider boron in epoxy resin with their elastic properties in Gpa: $G_f = 172$, $K_f = 228$ and $G_m = 1.53$, $K_m = 4.59$ [2]. The relative longitudinal (G_{13}/G_m) and transverse (G_{12}/G_m) shear moduli given by Eq. (7.21) are plotted in Figure 7.2 for $\phi = 0.2$ and 0.6. G_{12}/G_m decreases sharply with an increase in ρ from $\rho < 1$ and approaches its asymptote near $\rho = 10$. The variation of G_{13}/G_m is small but increases with particle concentration. The relative bulk modulus is computed in Figure 7.3, which has a minimum at $\rho = 1$, as expected. The relative longitudinal and transverse Young

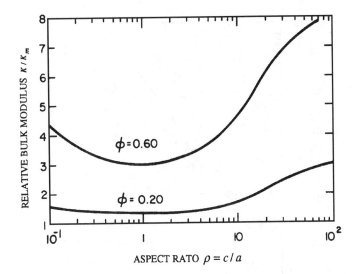

FIGURE 7.3. Dependence of the bulk modulus on the particle shape and volume fraction.

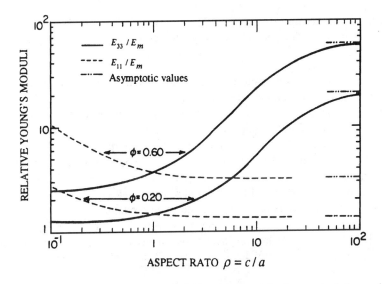

FIGURE 7.4. Dependence of Young's modulus on the particle shape, orientation and volume fraction.

moduli are calculated in Figure 7.4. It clearly illustrates the reinforcing effect of orienting fibrous filler ($\rho > 1$) parallel or disk ($\rho < 1$) perpendicular to the stretching direction for all values of ϕ. When $\rho = 1$, $G_{12} = G_{13}$, $E_{33} = E_{11}$, and all of the above calculations agree exactly with those obtained from Kerner's formula.

The effective longitudinal Young modulus, E_{33}, has received much attention because of the strong interest in the high-performance reinforced composites. For

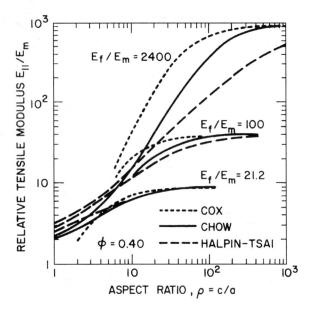

FIGURE 7.5. A comparison of Eq. (7.25) with other expressions for the longitudinal Young's modulus at different ratios of the elastic constants between the filler and matrix [10].

predicting E_{33} of a composite filled with aligned fibers randomly distributed in a matrix, several equations, including Eq. (7.25), are compared in Figure 7.5. The ratio of E_f/E_m ranges from 21.2 (glass in epoxy resin) [2] to 100 (boron in epoxy resin) [9] to 2400 (semicrystalline polyethylene) [9] at constant $\phi = 0.40$. The Cox equation [11] is limited for $\rho \gg 1$, because this model is known as the shear-lag model and does not taken into account the normal stresses in the matrix. The tensile stresses are carried by fibers and the load is transferred from fiber to fiber by shear. The Halpin-Tsai is an empirical equation [2]. For low values of E_f/E_m, little difference exists among the three predictions, especially for large aspect ratios. The discrepancy among them increases with increasing values of $E_f/E_m > 10^2$. Experimental verification is usually needed for aspect ratio other than one or infinity over a wide range of E_f/E_m for technology applications.

7.3   Thermal Expansion

The investigation of the effective thermal expansion of a particulate composite in terms of the phase geometry, configuration, and material properties of each individual constituent has been extensively reported in the literature [12]. The two purposes of this section are (1) to analyze the particle-shape effect on the anisotropy thermal expansion of composites filled with randomly distributed aligned spheroids at finite concentration, and (2) to illustrate the extension of an effective medium theory reported in the last section for the elastic moduli of two-phase systems.

When fillers are considered as identical spheroids given by Eq. (7.7) with corresponding axes aligned and the spatial distribution of the particles is random and homogeneous. The composite as a whole is transversely isotropic about the x_3 axis. The effective volumetric thermal expansion (γ) is related to the effective longitudinal (θ_{33}) and transverse ($\theta_{11} = \theta_{22}$) linear thermal expansion coefficients by

$$\gamma = 2\theta_{11} + \theta_{33}. \tag{7.27}$$

Because the linear thermal expansion coefficients of the filler (θ_f) and matrix (θ_m) are different, the internal strains

$$\varepsilon_{ij} = (\theta_f - \theta_m)\Delta T \delta_{ij} \tag{7.28}$$

are generated in the system in which ΔT is the temperature change. This process creates the local stresses:

$$\sigma_{pq}(\vec{r}) = C^{(m)}_{pqij} e^C_{ij}(\vec{r}) \quad \text{in } V_m \tag{7.29a}$$

and

$$\sigma_{pq}(\vec{r}) = C^{(f)}_{pqij}\left[e^C_{ij}(\vec{r}) - \varepsilon_{ij}\right] = C^{(m)}_{pqij}\left[e^C_{ij}(\vec{r}) - e^T_{ij}(\vec{r})\right] \quad \text{in } V_f. \tag{7.29b}$$

Particles are firmly bonded to the matrix.

For the macroscopically homogeneous composite, the volume average of local thermal stresses over the total volume has to be zero. Using Eq. (7.29), we obtain

$$\frac{1}{V}\int_V \sigma_{pq}(\vec{r})\, dV = \frac{1}{V}\left[\int_{V_m} \sigma_{pq}(\vec{r})\, dV + N\int_{V_f} \sigma_{pq}(\vec{r})\, dV\right]$$

$$= \frac{C^{(m)}_{pqij}}{V}\left\{\int_{V_m} e^C_{ij}(\vec{r})\, dV + N\int_{V_f}\left[e^C_{ij}(\vec{r}) - e^T_{ij}(\vec{r})\right] dV\right\} = 0. \tag{7.30}$$

By using the notations defined in Eq. (7.14), Eq. (7.30) becomes

$$(1 - \phi)\langle e^C_{ij}\rangle_m + \phi\left[\langle e^C_{ij}\rangle_f - \langle e^T_{ij}\rangle_f\right] = 0. \tag{7.31}$$

Let us again assume the average constrained strain tensor is related to the transformed strain tensor by Eq. (7.16) at the finite ϕ. Substituting Eq. (7.16) into Eq. (7.31), we get

$$\langle e^c_{ij}\rangle_f = (1 - \phi)S_{ijkl}\langle e^T_{kl}\rangle_f + \phi\langle e^T_{ij}\rangle_f. \tag{7.32}$$

This equation is for thermal expansion that differs from Eq. (7.17) for elastic constants. When both filler and matrix are isotropic, two independent equations are derived from eqs. (7.28), (7.29b), and (7.32):

$$2\tilde{B}_1\langle e^T_{11}\rangle_f + \tilde{B}_3\langle e^T_{33}\rangle_f = (K_f/K_m)(\gamma_f - \gamma_m)\Delta T \tag{7.33a}$$

and

$$\tilde{C}_1 \langle e_{11}^T \rangle_f = \tilde{C}_3 \langle e_{33}^T \rangle_f, \tag{7.33b}$$

where

$$\tilde{B}_i = 1 + (K_f/K_m - 1)[(1 - \phi)\vartheta_i + \phi], \qquad i = 1, 3$$

and

$$\tilde{C}_i = 1 + (G_f/G_m - 1)[(1 - \phi)\chi_i + \phi], \qquad i = 1, 3.$$

Both ϑ_i and χ_i are a function of the aspect ratio ρ and Poisson's ratio ν_m of the matrix [see Eq. (7.24) and Appendix 7A].

The effective linear thermal expansion coefficients of the anisotropy composite is expressed by the average transformation strains as [see Eq. (7.20)]:

$$\theta_{ii} = \theta_m + \langle e_{ii}^T \rangle_f \phi/\Delta T, \qquad ii = 11, 33. \tag{7.34}$$

Substituting the solution of Eq. (7.33) into Eq. (7.34), we obtain the effective longitudinal, linear thermal expansion coefficient

$$\theta_{33} = \theta_m + \frac{K_f}{K_m} \frac{(\gamma_f - \gamma_m)\tilde{C}_1 \phi}{2\tilde{B}_1 \tilde{C}_3 + \tilde{C}_1 \tilde{B}_3}, \tag{7.35}$$

and the effective transverse, linear thermal expansion coefficient

$$\theta_{11} = \theta_m + \frac{K_f}{K_m} \frac{(\gamma_f - \gamma_m)\tilde{C}_3 \phi}{2\tilde{B}_1 \tilde{C}_3 + \tilde{C}_1 \tilde{B}_3}. \tag{7.36}$$

Thus, the effective volumetric thermal expansion coefficient is

$$\gamma = \gamma_m + \frac{K_f}{K_m} \frac{(\gamma_f - \gamma_m)(\tilde{C}_1 + 2\tilde{C}_3)\phi}{2\tilde{B}_1 \tilde{C}_3 + \tilde{C}_1 \tilde{B}_3}. \tag{7.37}$$

When $\rho = 1$, $\theta_{11} = \theta_{33} = \theta$ and eqs. (7.35)–(7.37) reduce to

$$\gamma = 3\theta = \gamma_m + \frac{K_f}{K_m} \frac{(\gamma_f - \gamma_m)\phi}{1 + (K_f/K_m - 1)[(1 - \phi)\vartheta + \phi]}, \tag{7.38}$$

which is exactly the Kerner equation [8]. Comparing eqs. (7.38) and (7.24) at $\phi \to 0$, we get the relationship between the bulk modulus and the volumetric expansion:

$$\frac{\gamma - \gamma_m}{\gamma_f - \gamma_m} = \frac{K_f}{K_m} \frac{K - K_m}{K_f - K_m}, \tag{7.39}$$

which is valid for all values of ρ.

Consider glass in epoxy resin. Their shear and bulk moduli (in Gpa) are $G_f = 29.9$, $K_f = 43.5$; and $G_m = 1.28$, $K_m = 3.83$. The effective thermal expansion coefficients are calculated as a function of ϕ and ρ. The longitudinal and transverse, linear thermal expansion coefficients are shown in Figure 7.6. They show strong

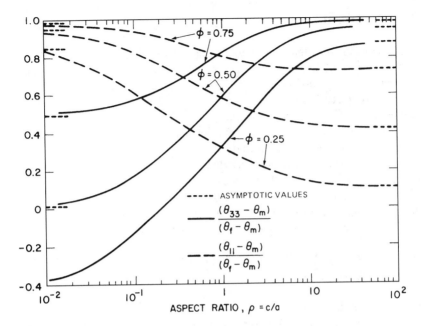

FIGURE 7.6.   The influence of the particle shape, orientation and volume fraction on the linear thermal expansion coefficient.

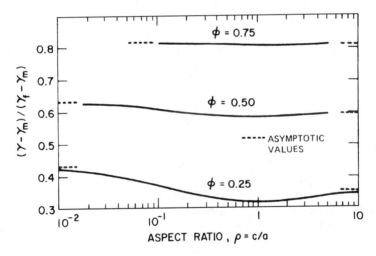

FIGURE 7.7.   The influence of the particle shape and volume fraction on the volumetric thermal expansion coefficient.

anisotropy dependence on the particle shape and approach their asymptotes at $\rho = 10$ and 10^{-2} for fibrous or disk-like particles, respectively. Figure 7.7 shows that the volumetric thermal expansion coefficient is not sensitive to the particle shape. The effect of particle shape diminishes as the particle concentration increases.

7.4   Stress Concentration

The ratio of the maximum stress to the applied stress defines the stress concentration factor. For aligned prolate spheroids [see Eq. (7.7) with $\rho = c/a > 1$], the maximum stress occurs at $(x_1, x_2, x_3) = (0, 0, c)$, where cohesive or adhesive failure may take place. Therefore, the determination of the stress concentration factor is important to heterogeneous materials.

Traditionally, the analysis of the stress concentration has been limited to a single isolated particle that does not include the concentration dependence. We include the particle–particle interaction here that follows exactly the same approach of using the effective medium theory discussed in the previous sections. The spatial average of the internal stress tensor, $\langle \sigma_{pq} \rangle_f$, inside prolate spheroids, caused by an uniform applied stress tensor σ_{ij}^A, can be derived in the same way as that for Eq. (7.19). The internal stress tensor is determined from

$$(1 - \phi)\lfloor(\lambda_f - \lambda_m)S_{kkpq}\delta_{ij} + 2(G_f - G_m)S_{ijpq}\rfloor\langle\sigma_{pq}\rangle_f + \lambda_m\langle\sigma\rangle_f + 2G_m\langle\sigma_{ij}\rangle_f$$
$$= \lambda_f\sigma^A\delta_{ij} + 2G_f\sigma_{ij}^A, \tag{7.40}$$

where σ is the stress invariant.

Consider a uniform tensile stress σ_{33}^A as the only nontrivial stress applied to the system. Solving Eq. (7.40) gives the stress concentration factor,

$$F_{sc} \equiv \frac{\langle\sigma_{33}\rangle_f}{e_{33}^A} = \frac{(K_f/K_m)C_1 + 2(G_f/G_m)B_1}{2B_1C_3 + C_1B_3}, \tag{7.41}$$

where B_i and C_i $(i = 1, 3)$ are given in Eq. (7.24). The above equation contains both particle shape and concentration effects. When $\phi \to 0$ and $\rho = 1$, Eq. (7.41) becomes

$$F_{sc} = \frac{1}{3}\left[\frac{K_f/K_m}{1 + (K_f/K_m - 1)\vartheta} + \frac{2G_f/G_m}{1 + (G_f/G_m - 1)\chi}\right], \tag{7.42}$$

where ϑ and χ are given in Appendix 7A. Eq. (7.42) is the classical Goodier equation [13]. For illustration, consider a glass-filled poly(phenylene oxide) possessing the following properties: $E_f = 73.1$ Gpa, $v_f = 0.22$; and $E_m = 2.60$ Gpa, $v_m = 0.35$. The stress concentration factor is calculated from Eq. (7.41) as a function of ρ and ϕ. The general behavior is shown in Figure 7.8. It reveals that the concentration dependence diminishes for large values of the aspect ratio.

7.5   Nonequilibrium Mechanical Properties

The change in the physical mechanism of deformation from elasticity to viscoelasticty to plasticity depends on the time scales in which the composite system are measured and relaxed. Chapters 5 and 6 have shown that segment mobility

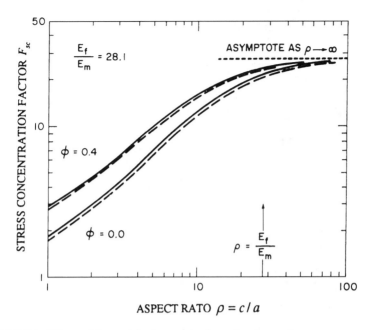

FIGURE 7.8. Effects of the particle shape and volume fraction on the stress concentration factor.

and structural relaxation in amorphous polymers play an important role in determining the yield behavior that is closely related to the nonlinear viscoelastic phenomenon. The influence of structural relaxation will now be incorporated into the equilibrium composite theory discussed in previous sections for the purpose of describing the nonequilibrium mechanical properties of polymer composites. A pertinent rule of mixtures for the compositional-dependent relaxation time has to be introduced. A unified theory will then be presented that enable us to predict not only the elastic constants, but also the yield stress and the nonlinear stress–strain behavior.

Let us consider the composite as a disordered solid consisting of randomly distributed spherical particles and holes in a polymer matrix under uniaxial compression. The interaction between particles has been treated by an effective medium theory. The effective Young modulus is given by [see Eq. (7.25)]

$$\frac{E(\phi)}{E_m} = 1 + \frac{1}{3}\left[\frac{K_f/K_m - 1}{1 + (K_f/K_m - 1)(1 - \phi)\vartheta} + \frac{2(G_f/G_m - 1)}{1 + (G_f/G_m - 1)(1 - \phi)\chi}\right]\phi,$$
(7.43)

where ϑ and χ are given in Appendix 7A. The use of this equation is shown in Figure 7.9, where the calculated and measured Young's modulus of crosslinked epoxy resins filled with silica (SiO_2) is compared. The elastic properties of the filler and matrix are $E_f/E_m = 21.2$, $v_f = 0.22$; and $E_m = 17 \times 10^3$ kg/cm^2, $v_m = 0.35$.

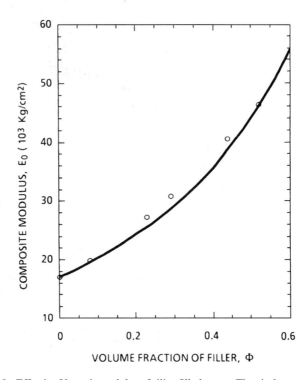

FIGURE 7.9. Effective Young's modulus of silica filled epoxy. The circles are experimental data [14].

In order to understand the deformation kinetics of a composite, one need to know the pertinent rule of mixtures that defines the compositional dependent of relaxation time. Consider the lattices for binary mixtures that consist of the number of lattice sites [see Eq. (5.1)]

$$N_j(t) = n_j(t) + x_j n_{xj}, \qquad j = f, m. \tag{7.44}$$

Each lattice site occupies a single lattice volume v_j. Blends of two-phase materials are expected to exhibit no volumetric deviation from an additive relationship:

$$V = v_f N_f + v_m N_m \equiv vN = v(n + xN). \tag{7.45}$$

This process results in the effective number of holes

$$n = \frac{v_f n_f + v_m n_m}{v}$$

and the free-volume fraction of the composite

$$f = n/N = f_m + (f_f - f_m)\phi, \tag{7.46}$$

where $f_j = n_j/N$ and $\phi = v_f N_f/vN$.

In single-phase material, the relaxation time is related to the hole fraction by Eq. (5.44). By using Eq. (7.46), the relaxation time can written as

$$\ln\left[\frac{\tau(\phi)}{\tau_m}\right] = \frac{b}{f} - \frac{b_m}{f_m} = \frac{(b - b_m) - b_m(f_f/f_m - 1)\phi}{f_m(1 - \phi) + f_f\phi}, \qquad (7.47)$$

where b and b_m are constants. Because the high-modulus filler is much denser than the polymer matrix, we may assume $f_f/f_m \to 0$ [15] and obtain the composite relaxation time

$$\log\left[\frac{\tau(\phi)}{\tau_m}\right] = \frac{1}{2.303}\ln(a_\phi) = \frac{c\phi}{1 - \phi}, \qquad (7.48)$$

where $c = b/2.303 f_m$ and a_ϕ is the concentration-dependent shift factor. Eq. (7.48) is obtained for a disordered solid that contains not only filler particles, but also holes in the polymer matrix. The volume of the system is closely packed, and no interpenetrating of chain molecules and the hole occurs at the interface between the particle and polymer.

Substituting Eq. (7.48) into Eq. (6.31) gives the compressive yield stress

$$\sigma_y = Z + J\left(\log\dot{e} + \frac{c\phi}{1 - \phi}\right), \qquad (7.49)$$

where $J \sim \Omega_{11}^{-1}$. Both Z and J are constants. A comparison of Eq. (7.49) with experimental data at room temperature (23 °C) is shown in Figure 7.10, where the

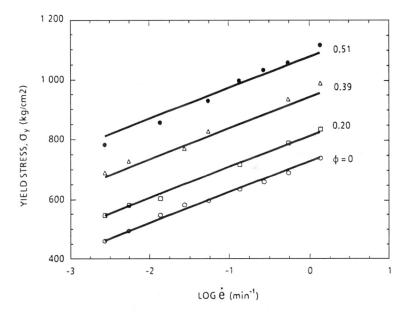

FIGURE 7.10. Comparison of the predicted (solid lines) and measured (points [14]) strain-rate dependence of the compressive yield stress of silica filled epoxy at different filler concentrations.

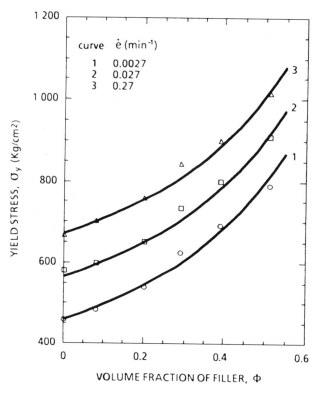

FIGURE 7.11. The compressive yield stress of silica filled epoxy versus filler concentration at different strain rates. The points are experimental data [14].

compressive yield stress is plotted against the strain rate at different filler contents. The slope gives the value of J, and it reveals that the activation volume is not a function of ϕ. This process supports the notion that the deformation kinetics for a composite is mainly defined by the structural relaxation in the polymer matrix. The constant c defines the plots of the yield stress versus filler concentration in Figure 7.11. These two figures give $J = 105$ kg/cm^2 and $c = 3.2$ for the silica-filled epoxy resins. Experimental data used in figures 7.9, 7.10, and 7.11 are taken from the same source.

By following Eq. (6.60), the nonlinear stress–strain behavior is determined by

$$\sigma(e) = E(\phi) \int_o^e \exp\left\{-\left[\frac{e' \exp(2.303\sigma(e')/J)}{\dot{e}\tau_m a_\phi}\right]^\beta\right\} de'. \qquad (7.50)$$

The compositional-dependent, effective Young modulus, $E(\phi)$, and shift factor, a_ϕ are given by eqs. (7.43) and (7.48). The parameters $\beta = 0.19$ and $\tau_m = 7.11 \times 10^{12}$ sec at $23\,^\circ$C are adopted in seeking the numerical solution of Eq. (7.50). As mentioned earlier, β for the composite system has the same value as that for epoxy resins (see Section 6.3) because the distribution of relaxation times

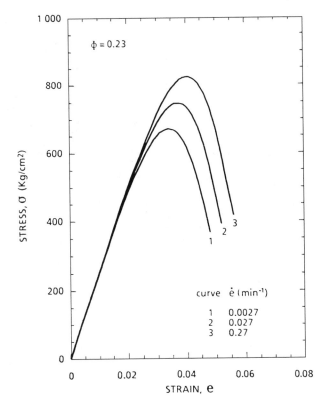

FIGURE 7.12. The compression stress–strain relationships of silica filled epoxy at different strain rates.

may not be affected by the presence of strong fillers. The effect of strain rate is shown in Figure 7.12. The nonlinear compressive stress–strain curves at different filler concentrations are plotted in Figure 7.13. As the volume fraction of filler increases, the yield stress of composite system increases, but at the same time the system becomes more brittle.

7.6   Compatible Polymer Blends

In contrast to two-phase composites, the miscible glassy blends have shown an unusual maximum yield stress at a critical concentration. Blending of polymers has received lots of attention as a method of improving the physical properties of polymers. The Flory–Huggins interaction parameter has played a key role in the thermodynamic investigation of mixing in the liquid state [16]. Little development has occurred, however, in the understanding of the interactions in the glassy state that could help us to explain the stress anomaly in the compatible blends. We intend to look into the role of the excess volume of mixing for $T < T_g$ and then use it to explain this stress anomaly.

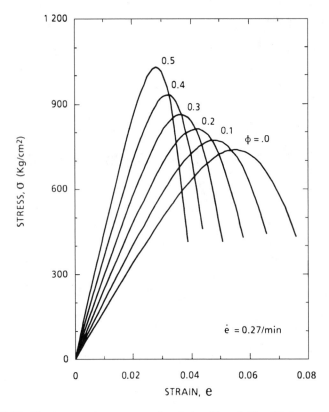

FIGURE 7.13. The compression stress–strain relationships of silica filled epoxy at different filler concentrations.

In accordance with Eq. (7.44), the total volume of a compatible blend is

$$V = \sum_{j=1}^{2} v_j N_j + \Delta V_m \equiv vN = v(n + x n_x). \tag{7.51}$$

The excess volume of mixture ΔV_m can be written in the form

$$\Delta V_m / vN = A\phi(1 - \phi) \leq 0, \tag{7.52}$$

where the volume fraction $\phi \equiv \phi_1 = 1 - \phi_2$. The nondimensional parameter A measures the strength of the volume interaction between the two components in the blend for $T < T_g$. Because the volumes of polymer chains should remain unchanged before and after mixing, that means

$$v x n_x = \sum_{j=1}^{2} v_j x_j n_{xj}. \tag{7.53}$$

Subtracting Eq. (7.53) from Eq. (7.51) and using Eq. (7.52), we obtain the free-volume fraction of the blend

$$f = \frac{vn}{vN} = \sum_{j=1}^{2} \phi_j f_j + A\phi(1 - \phi). \tag{7.54}$$

The value of A can be determined experimentally by using the more explicit expressions of eqs. (7.51) and (7.52):

$$V = [V_2 + (V_1 - V_2)\phi][1 + A\phi(1 - \phi)]. \tag{7.55}$$

This equation is compared with experimental data on blends of poly(2,6-dimethyl-1,4-phenylene oxide) (PPO) and poly(styrene-co-p-chlorostyrene) (PS) in Figure 7.14. We find $A = -0.034$ at $T = 20\,°C$ and $100\,°C$ (both temperatures are below T_g). The value of A is clearly independent of temperature in the glassy state; however, measured data for $T \geq T_g$ [18] have shown that $A = 0$.

The glass transition temperature of the blends (T_g) is not a constant but a function of ϕ. Let us look at Eq. (7.54) with $A = 0$ at $T = T_g$. The equilibrium free-volume fractions are related to the hole energies ε_j by Eq. (5.2). Assuming $f_{r1} = f_{r2} = f_r$

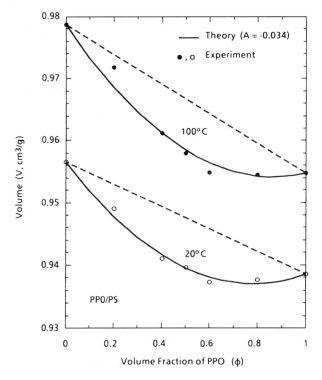

FIGURE 7.14. Comparison of the calculated and measured [17] volume of PPO/PS blends as a function of the volume fraction of PPO at ambient pressure.

to be a universal constant (see Section 5.1), we obtain

$$\sum_{j=1}^{2} \phi_j \exp\left[\frac{\varepsilon_j}{k}\left(\frac{1}{T_g} - \frac{1}{T_{gj}}\right)\right] = 1. \tag{7.56}$$

As a first approximation, Eq. (7.56) gives

$$\frac{1}{T_g} = \sum_{j=1}^{2} \frac{\varepsilon_j \phi_j}{T_{gj}} \bigg/ \sum_{j=1}^{2} \varepsilon_j \phi_j \tag{7.57}$$

and

$$\frac{T_g}{T_{g2}} = \frac{1 + (\varepsilon_1/\varepsilon_2 - 1)\phi}{1 + [(\varepsilon_1/\varepsilon_2)(T_{g2}/T_{g1}) - 1]\phi}. \tag{7.58}$$

Fitting Eq. (7.58) with experimental data of PPO/PS blends in Figure 7.15, we get $\varepsilon(PPO)/\varepsilon(PS) = 0.92$.

The yield stress of glassy polymer is closely related to the effective relaxation time of compatible blends [see Eq. (7.47)]:

$$\ln\left[\frac{\tau(\phi)}{\tau_2}\right] = \frac{b}{f} - \frac{b_2}{f_2} = \frac{(b - b_2) - b_2(f_1/f_2 - 1)\phi}{f}, \tag{7.59}$$

where b and b_2 are constants and f is given by Eq. (7.54). When $\phi \to 0$, one gets $\tau/\tau_2 \to 1$, which requires $b = b_2$. By introducing $c = b/2.303 f_2$ and the intrinsic

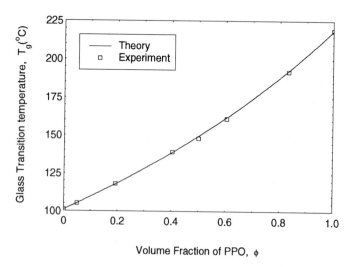

FIGURE 7.15. A comparison of the calculated and measured [18] glass transition temperature of PPO/PS blends as a function of the volume fraction of PPO.

hole fraction

$$[f] \equiv \frac{f - f_2}{f_2 \phi} = \left(\frac{f_1}{f_2} - 1\right) + \frac{A}{f_2}(1 - \phi), \qquad (7.60)$$

Combining eqs. (7.59) and (7.60) gives

$$\log\left[\frac{\tau(\phi)}{\tau_2}\right] = -\frac{c[f]\phi}{1 + [f]\phi}. \qquad (7.61)$$

In the case of two-phase composites ($A = 0$ and $f_1 = 0$), we get $[f] = -1$ and Eq. (7.61) reduces to Eq. (7.48). From Eq. (7.61), we get [see eqs. (6.31) and (7.49)]

$$\sigma_y = Z + J\left(\log \dot{e} - \frac{c[f]\phi}{1 + [f]\phi}\right). \qquad (7.62)$$

For a fixed strain rate, a comparison of Eq. (7.62) and experimental data of miscible blends is shown in Figure 7.16. Curves 1 and 2 represent the PPO/PS blends

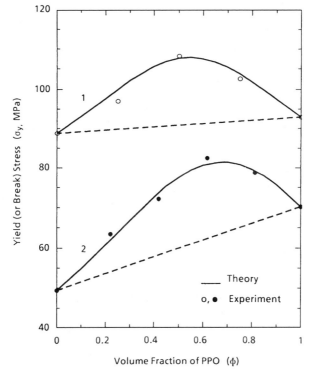

FIGURE 7.16. Comparison of the calculated and measured yield stress. Curve 1 is for the PPO/PS blends under uniaxial compression [19], and curve 2 is for the PPO/PS-pCIS blends under uniaxial tension [20].

TABLE 7.1. *Physical parameter of polymer blends [21]*

Parameter	1: PPO/PS in Compression	2: PPO/PS-pCIS* in Tension
A	−0.034	
A/f_2	−1.40	−1.00
f_1/f_2	0.87	0.64
cJ, MPa	26.7	37.2

*The blends of PPO and a random copolymer with 58.6 mole%
of pCIS (p-chlorostyrene)

in compression and the PPO/PS–pCIS blends in tension, respectively. Table 7.1 lists the three parameters used in Figure 7.16. The unexpected feature here is the presence of the maximum yield stresses between $0 < \phi < 1$. It is a result of the negative nonequilibrium interaction ($A < 0$). We prefer to have $-A/f_2 > 1$ and a larger difference between the yield stresses of blending polymers in the search of strong polymer blends. The presence of a maximum yield stress at a critical concentration is a unique feature of compatible blends, and it does not occur in multiphase composites.

7.7 Molecular Composites

We have seen that the glass transition temperature is an important physical property in the discussion of molecular composites in the last section. The glassy state represents a situation of frozen-in disorder that is in a state of quasi-equilibrium, and the glass transition can be explained by using either equilibrium thermodynamics or kinetics (see Section 6.1). The approach here is the equilibrium one because it is more convenient to derive an expression of T_g by the molecular weight, size, and concentration of small molecules.

According to the view of GD (see Section 6.1), glasses are formed as a result of a system losing its configuration entropy: $S_c = S^{liquid} - S^{glass}$. For pure polymers, $S_c(0, T)$ is related to the heat capacity at constant pressure (C_p) by

$$S_c(0, T) = \int_{T_{g0}}^{T} \Delta C_p(T') d \ln T', \qquad (7.63)$$

where T_{g0} is the glass transition temperature of pure polymer and ΔC_p is the difference in heat capacity between the supercooled liquid and glass. The configuration entropy of a polymer-small molecule system also depends on the number of small molecules (ℓ) that has

$$S_c(\ell, T) = \int_{T_g}^{T} \Delta C_p(\ell, T') d \ln T'. \qquad (7.64)$$

When the energy contributed from the vibration about the lattice sites is neglected, we have mentioned in Section 6.1 that the entropy $S^{glass}(0, T)$ goes to zero at the transition temperature T_2, which is about $51\,^{\circ}C$ below the glass transition temperature [Ref. 1 in Chapter 6] and not easily reachable. Nevertheless, the composition dependence of T_g and T_2 is assumed to be the same. By setting $S^{glass}(\ell, T) = S^{glass}(0, T) = 0$ at the glass transition and approximating ΔC_p to be independent of temperature and composition, eqs. (7.63) and (7.64) lead to

$$\ln\left(\frac{T_g}{T_{g0}}\right) = -\frac{1}{\Delta C_p}[S^{liquid}(\ell, T) - S^{liquid}(0, T)]. \tag{7.65}$$

In general, the entropy is related to the configuration partition (Q) by

$$S = k \ln Q + kT \left(\frac{\partial \ln Q}{\partial T}\right)_p. \tag{7.66}$$

Substituting Eq. (7.66) into Eq. (7.65) gives

$$\ln\left(\frac{T_g}{T_{g0}}\right) = -\frac{1}{\Delta C_p}\left[\ln\left(\frac{Q_\ell^{liquid}}{Q_0^{liquid}}\right) + T\frac{\partial}{\partial T}\ln\left(\frac{Q_\ell^{liquid}}{Q_0^{liquid}}\right)\right]. \tag{7.67}$$

Any one of the rules of mixture may be used to evaluate the ratio of the partition functions. In the problem of plasticizing a high polymer by small molecules, the first-order interest is to determine the possible configurations of small molecules on polymer lattice sites rather than the rearrangements of polymer molecules because the glass transition of a pure polymer has already been assumed to be known. To find the ratio of T_g / T_{g0}, the main contribution to the partition function is from the mixing of the small molecules on lattice sites.

Consider ℓ small molecules randomly distributed in a lattice of $\ell + L$ sites, where L is the number of vacant lattice sites. Let us assume that each small molecule occupies a single lattice site. We write [22]

$$Q_0^{liquid} = 1 \tag{7.68}$$

and

$$Q_\ell^{liquid} = \frac{(\ell + L)!}{\ell! L!}\exp\left(\frac{\ell L}{\ell + L}\frac{z\varepsilon}{2kT}\right), \tag{7.69}$$

where z is the lattice coordinate number, $\varepsilon = \varepsilon_{\ell\ell} + \varepsilon_{LL} - 2\varepsilon_{\ell L}$. Here, $\varepsilon_{\ell\ell}$, ε_{LL}, and $\varepsilon_{\ell L}$ are energies of each $\ell\ell$, LL, and ℓL pair. Substituting eqs. (7.68) and (7.69) into Eq. (7.67), we obtain

$$\ln\left(\frac{T_g}{T_{g0}}\right) = \frac{zR}{M_p \Delta C_p}[(1 - \theta)\ln(1 - \theta) + \theta \ln \theta], \tag{7.70}$$

where R is the gas constant and $\theta = \ell/(\ell + L)$ is a nondimensional parameter. In practice, we write $\ell = m_s N_A / M_s$ and $\ell + L = z m_p N_A / M_p$, where N_A is Avogadro's number, m is the mass, and M is the molecular weight. The subscripts

s and p identify the small molecule and polymer, respectively. M_p is the monomer molecular weight. Thus,

$$\theta = \frac{M_p}{zM_s}\frac{w}{1-w} = \frac{V_p}{zV_s}\frac{\phi}{1-\phi}, \qquad (7.71)$$

where w is the weight fraction of the small molecule. The corresponding volume fraction is ϕ, and V is the molar volume. Eq. (7.70) predicts $T_g(\phi)$ by the molecular parameters and has the advantage of not requiring the knowledge of the glass transition temperature of small molecules, which cannot be determined easily.

In the derivation of Eq. (7.70), θ was implicitly assumed to be small; however, it does not limit its usefulness. As an example, let us consider the many different small molecules being added to polystyrene, which has the following properties: $M_p = 104.15$ g/mol, $M_p\Delta C_p = 6.45$ cal/mol $- K$, $z = 2$, and $T_{g0} = 358.5$ K. The depression of the glass transition temperature of polystyrene by the small molecules of different molecular weight, size (molar volume), and concentration is shown in Figure 7.17. The solid curve represents theoretical calculation, and the symbols for experimental points are given in Table 7.2. Clearly, $T_g(\phi)$ decreases as

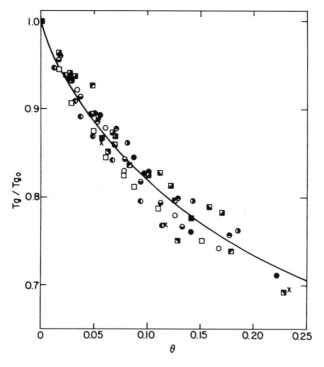

FIGURE 7.17. The depression of the glass transition temperature of polystyrene by small molecules of different molecular weight, size and concentration. Theory (solid curve) is compared with experiment (points [23]). The properties of the data points are listed in Table 2.

TABLE 7.2. *Various small molecules in polystyrene*

Molecules	M_s $\dfrac{g}{mol}$	V_s $\dfrac{cm^3}{mol}$	Symbols in Figure 7.17
methyl acetate	74.08	79.30	○
CS_2	76.14	60.28	◖
C_6H_6	78.12	88.87	◗
CH_2Cl_2	84.93	64.20	◑
ethyl acetate	88.11	98.34	◐
$C_6H_5CH_3$	92.14	106.03	●
n-butyl acetate	116.16	131.70	□
$CHCl_3$	119.38	80.94	◪
$C_6H_5NO_2$	123.11	102.28	◪
methyl salicylate	152.15	128.95	◩
CCl_4	153.82	96.49	◪
phenyl salicylate	214.22	171.38	×
β-naphthyl salicylate	264. 22	214.00	■

the molecular weight, size, and concentration of small molecule increase. Eq. (7.70) applies well to molecular composites containing a low content of small molecule to the high concentration of larger molecules.

7.8   Nanocomposites

The focus here is on the microscopic heterogeneity in composites with structures ranging from a fraction of nanometer to less than a micron. The molecular composites discussed in the last section are also within this size range. We would like to know in this section what is the effect of small molecules on the effective elastic constant of nanocomposites. This knowledge enables us to see the limitation of the traditional composite theories, which are based on micromechanics, and point out the importance of molecular parameters. It is reasonable to start the investigation of the composite theory presented in Section 7.2 for the effective elastic constants on the basis of continuum mechanics, in which the elastic energy is conserved. On the other hand, the properties such as the glass transition and structural relaxation originated from the change in entropy caused by the introduction of small molecules. We shall examine the physics behind the breakdown of the composite theory in its application down to the nano-size range. The crossover from the energetic to entropy contributions to the elastic constants will be discussed.

When a solid is deformed under a stress tensor σ_{ij}, the change in deformation is expressed by the strain tensor e_{ij}. The work done by the internal stresses per unit volume of the body is then

$$dW = -\sum_{ij} \sigma_{ij} de_{ij}. \tag{7.72}$$

It is assumed that the process of deformation occurs slowly and thermodynamic equilibrium is established in the system at every instant. The combined equation of the first and second laws of thermodynamics is

$$TdS = dU + dW = dU - \sum_{ij} \sigma_{ij} de_{ij}.$$ (7.73)

The free energy of the system is $F = U - TS$. Competition occurs between the energy U (which favors order) and entropy (which favors disorder). Under a constant temperature, a rise in entropy caused by the presence of small molecules in a polymer tilts the balance toward disorder (see Section 7.7). Using Eq. (7.73), we obtain the differential of F to be

$$dF = -SdT + \sum_{ij} \sigma_{ij} de_{ij}.$$ (7.74)

It follows that $\sigma_{ij} = (\partial F/\partial e_{ij})_{Te'}$. The subscript e' means that all other e_{ij} is to hold constant while differentiating with respect to e_{ij}. The fourth-order tensor of elastic constants is determined from [6,24]

$$C_{ijkl} = \left(\frac{\partial \sigma_{ij}}{\partial e_{kl}}\right)_{e'} = \left(\frac{\partial^2 F}{\partial e_{ij}\partial e_{kl}}\right)_{Te'} = \left(\frac{\partial^2 U}{\partial e_{ij}\partial e_{kl}}\right)_{Te'} - T\left(\frac{\partial^2 S}{\partial e_{ij}\partial e_{kl}}\right)_{Te'}.$$ (7.75)

The Maxwell equations are obtained by differentiating the dependent variables and by comparing the various derivatives in the form of second derivatives of the state functions. Thus, Eq. (7.74) gives

$$\left(\frac{\partial S}{\partial e_{ij}}\right)_{Te'} = -\left(\frac{\partial \sigma ij}{\partial T}\right)_e.$$ (7.76)

Substituting Eq. (7.76) into Eq. (7.75) yields

$$C_{ijkl} = C_{ijkl}^{(U)} + T\left(\frac{\partial C_{ijkl}}{\partial T}\right)_e.$$ (7.77)

For simplicity, let us consider the tensile (Young's) modulus:

$$E(\phi) = E_U + T\frac{\partial E(\phi)}{\partial T},$$ (7.78)

where ϕ is the volume fraction of small molecules. The subscript U in Eq. (7.78) and the superscript in Eq. (7.77) are identified as energy-related elastic constants. The second term in the above equation is related to the disorder caused by the presence of small molecules in polymer. We have also dropped the subscript e on the partial derivative in Eq. (7.78).

7.8.A   Energy Contribution

The effective tensile modulus, $E_U(\phi)$, in Eq. (7.78) can be determined from Eq. (7.25). The small molecule–polymer system can be considered as a disordered solid consisting of randomly distributed liquid-like spherical particles in a solid matrix. The shear modulus of the small molecule (G_s) is always many orders of magnitude smaller than that of the solid polymer (G_p), but the bulk modulus of the small molecule (K_s) and of the polymer (K_p) are comparable. Thus, $G_s = 0$ and $K_s \neq 0$ are assumed. The effective tensile modulus of the system is

$$\frac{E_U(\phi)}{E_p} = 1 - \frac{1}{3}\left[\frac{1 - K_s/K_p}{1 - (1 - K_s/K_p)(1 - \phi)\vartheta} + \frac{2}{1 - (1 - \phi)\chi}\right]\phi, \quad (7.79)$$

where ϑ and χ are given in Appendix 7A with the Poisson ratio v_m being replaced by v_p of the polymer.

7.8.B   Entropy Contribution

The tensile modulus drops more sharply for temperature beyond the glass transition. Figure 7.17 has already suggested that small molecules in the polymer serve as softeners. Following Eq. (7.78), we introduce the softening temperature [25], $T_s = -E_U/(\partial E/\partial T) > 0$, and rewrite Eq. (7.78) as

$$E = -\frac{\partial E}{\partial T}(T_s - T). \quad (7.80)$$

The softening temperature is directly proportional to the glass transition, $T_s(\phi) = T_g(\phi) + d$, where $d > 0$ is a constant. The concept of T_s has found its usefulness in describing other properties like yield stress, $\sigma_y \sim T_s - T$ (see Figure 6.14). T_s is a signature of the onset of large-scale motion of polymer chains that results in the vanishing elastic constants and yield stresses.

Although the presence of small molecules in polymer has a strong effect on the configuration entropy and $T_g(\phi)$ given by Eq. (7.70), its influence on the tensile modulus has resulted in the change in the internal strain energy. We expect a crossover from the strain energy dominated to configuration entropy dominated $E(\phi)$ at a critical concentration ϕ_c. It is defined at $T_s(\phi) = T_g(0)$, which results in

$$d = T_g(0) - T_g(\phi_c). \quad (7.81)$$

For $\phi > \phi_c$, the mechanical properties are no longer determined by the local rearrangement of structure. The large-scale polymer motion related to the entropy change becomes important. Because ϕ is usually small in most applications, it is reasonable for us to assume that $\partial E(\phi)/\partial T \approx \partial E(\phi_c)/\partial T$. Together, with Eq. (7.80), this application leads to

$$\frac{E(\phi)}{E_U(\phi)} = \frac{T_s(\phi) - T}{T_s(\phi_c) - T}, \quad \text{for} \quad \phi \geq \phi_c. \quad (7.82)$$

Because $T < T_s(\phi) < T_s(\phi_c)$, the tensile modulus $E(\phi)$ is always less than $E_U(\phi_c)$

for $\phi > \phi_c$. Finally, we get

$$E(\phi) = E_U(\phi)\psi(\phi, T),\tag{7.83}$$

where

$$\psi(\phi, T) = 1, \quad \text{for } \phi \le \phi_c\tag{7.84a}$$

and

$$\psi(\phi, T) = \frac{T_g(\phi) + d - T}{T_g(0) - T}, \quad \text{for } \phi \ge \phi_c.\tag{7.84b}$$

Here, $E_U(\phi)$ is given by Eq. (7.79), $T_g(\phi)$ by eqs. (7.70) and (7.71), and the constant d by Eq. (7.81). The lowering of $E(\phi)$ and $T_g(\phi)$ is a measure of the size effect. When the size of the small molecule is small, it is more likely for it to interact with polymer segments and therefore alter the configuration entropy. This study clearly reveals that the ratio of V_p/zV_s is a measure of the nano-size effect.

As an example for illustration, we consider the cellulose-H_2O system. The lowering of the glass transition temperature is shown in Figure 7.18. The values of $M_p = 162$, $M_s = 18$, $z = 4$, $T_{g0} = 493$ K, and $M_p\Delta C_p = 10.48$ cal/mol$-K$ are used in the calculation of the solid curve from eqs. (7.70) and (7.71). The only adjusting parameter is z. On the basis of Eq. (7.83) and (7.84), a comparison

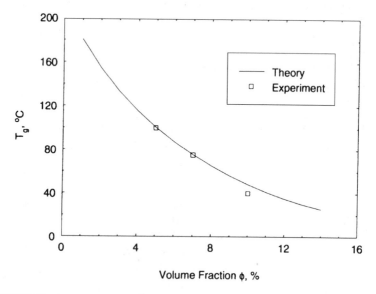

FIGURE 7.18. A comparison of the calculated and measured [26] depression of the glass transition temperature of cellulose by moisture.

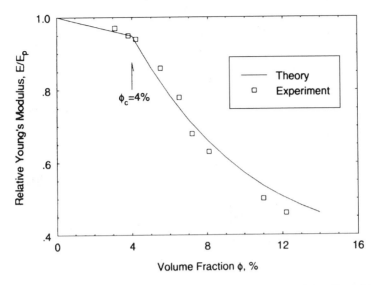

FIGURE 7.19. A comparison of the calculated and measured [26] tensile (Young's) modulus of cellulose-H_2O system at room temperature.

between the theoretical and experimental tensile modulus at $T = 296$ K is illustrated in Figure 7.19. The bulk properties [27]: $K_s = 2.22$ GPa, $K_p = 2.76$ GPa, and $v_p = 1/3$ are also needed in the calculation of $E(\phi)$. For $\phi < \phi_c$, the tensile modulus changes very little by following the traditional composite theory, Eq. (7.79). Beyond ϕ_c, the modulus decreases rapidly. The order–disorder transition occurs at $\phi_c = 0.04$. No such transition occurs in Figure 7.18 because the entropy alone determines the glass transition. Figure 7.19 clearly reveals how the traditional composite theory breaks down and points out the important role of molecular parameters in nanocomposites.

Appendix 7A   Eshelby's Tensor

$$S_{ijkl} = \frac{1}{2} \int_{V_1} C^{(m)}_{klpq} (g_{pi,qj} + g_{pj,qi}) \, dV, \qquad (7A\text{-}1)$$

where g is Green's function, which can be obtained in the similar fashion as Eq. (4.16) by solving the equation of equilibrium together with the linear constitutive equation. For a spherical particle ($\rho = 1$), Eq. (7A-1) reduces to

$$S_{ijkl} = \frac{1}{3}(\vartheta - \chi)\delta_{ij}\delta_{kl} + \frac{1}{2}\chi(\delta_{ik}\delta_{jl} + \delta_{il}\delta_{jk}), \qquad (7A\text{-}2)$$

where $\vartheta = 3 - 5\chi$ and χ is defined in Eq. (7.24). The pertinent parameters for spheroids are listed in the following:

$$\vartheta_i = \sum_{l=1}^{3} S_{llii}, \quad i = 1, 3,$$

$$\vartheta_1 = 4\pi Q/3 - 2(2\pi - I)R,$$

$$\vartheta_3 = 4\pi Q/3 - 4(I - \pi)R,$$

$$\chi_1 = S_{1111} + S_{1122} - 2S_{3311} = \left(\frac{4\pi}{3} - \frac{4\pi - 3I}{1 - \rho^2}\right)Q - 4(I - 2\pi)R,$$

$$\chi_3 = S_{3333} - S_{1133} = \left(\frac{4\pi}{3} - \frac{(4\pi - 3I)\rho^2}{1 - \rho^2}\right)Q + (4\pi - I)R,$$

$$2S_{1212} = \frac{2}{3}\left(\pi - \frac{1}{4}\frac{4\pi - 3I}{1 - \rho^2}\right)Q + 2IR,$$

$$2S_{1313} = \frac{1 + \rho^2}{3}\frac{4\pi - 3I}{1 - \rho^2}Q + (4\pi - I)R,$$

$$S_{1111} - S_{3311} = \left(\pi - \frac{7}{12}\frac{4\pi - 3I}{1 - \rho^2}\right)Q + (4\pi - I)R,$$

where

$$Q = \frac{3}{8\pi}\frac{1}{1 - v_m}, \qquad R = \frac{1}{8\pi}\frac{1 - 2v_m}{1 - v_m},$$

and

$$I = \frac{2\pi\rho}{(1 - \rho^2)^{3/2}}[\cos^{-1}\rho - \rho(1 - \rho^2)^{1/2}], \quad \text{for } \rho < 1,$$

$$I = \frac{2\pi\rho}{(\rho^2 - 1)^{3/2}}[\rho(\rho^2 - 1)^{1/2} - \cosh^{-1}\rho], \quad \text{for } \rho < 1.$$

When $\rho = 1$,

$$I = 4\pi/3 \quad \text{and} \quad \frac{4\pi - 3I}{1 - \rho^2} = \frac{4\pi}{5},$$

we obtain

$$\vartheta = \vartheta_1 = \vartheta_3 = \frac{1}{3}\frac{1 + v_m}{1 - v_m}$$

and

$$\chi = \chi_1 = \chi_3 = 2S_{1212} = 2S_{1313} = S_{1111} - S_{3311} = \frac{2}{15}\frac{4 - 5v_m}{1 - v_m}.$$

References

1. B. Sedlacek (Ed.), *Polymer Composites* (Walter de Gruyter, Berlin, 1986).
2. J. E. Ashton, J. C. Halpin, and P. H. Petit, *Primer on Composite Materials: Analysis* (Technomic, Stanford, CT, 1969).
3. J. L. Kardos, Crit. Rev. Solid State Sci. **3**, 417 (1973).
4. J. D. Eshelby, Proc. R. Soc. (London) A **241**, 376 (1957).
5. J. D. Eshelby, Proc. R. Soc. (London) A **252**, 561 (1959).
6. L. D. Landau and E. M. Lifshitz, *Theory of Elasticity* (Pergamon, Oxford, 1959).
7. T. S. Chow, J. Polym. Sci.: Poym. Phys. Ed. **16**, 959, 967 (1978).
8. E. H. Kerner, Proc. Phys. Soc. **69B**, 808 (1956).
9. W. T. Mead and R. S. Porter, J. Appl. Phys. **47**, 4278 (1976).
10. T. S. Chow, J. Mater. Sci. **15**, 1873 (1980).
11. H. L. Cox, British. J. Appl. Phys. **3**, 72 (1952).
12. I. Holiday and J. Robinson, J. Mater. Sci. **8**, 301 (1973).
13. J. N. Goodier, J. Appl. Mech. **1**, 39 (1933).
14. O. Ishai and L. J. Cohen, J. Composite Mater. **2**, 302 (1968).
15. T. S. Chow, Polymer **32**, 29 (1991)
16. D. R. Paul and S. Newman (Ed.), *Polymer Blends*, vol. 1 (Academic, New York, 1978).
17. P. Zoller and H. H, Hoehn, J. Polym. Sci.: Polym. Phys. Ed. **20**, 1385 (1982).
18. W. M. Prest, Jr., and R. S. Porter, J. Polym. Sci.: Polym. Phys. Ed. **10**, 1639 (1972).
19. R. P. Kambour and S. A. Smith, J. Polym. Sci.: Polym. Phys. Ed. **20**, 2069 (1982).
20. J. R. Fried, W. J. Macknight, and F. E. Karasz, J. Appl. Phys. **50**, 6052 (1979).
21. T. S. Chow, Macromolecules **23**, 4648 (1990).
22. T. S. Chow, Macromolecules **13**, 362 (1980).
23. E. Jenckel and R. Heusch, Kolloid-Z **130**, 80 (1953).
24. J. H. Weiner, *Statistical Mechanics of Elasticity* (Wiley, New York, 1983).
25. T. S. Chow, Macromolecules **26**, 5049 (1993).
26. N. L. Salmen and E. L. Back, Tappi **63**, 117 (1980).
27. D. E. Gray (Ed.), *American Institute of Physics Handbook*, 2nd ed. (McGraw-Hill, New York, 1963).

8

Rough Surfaces and Interfaces

All surfaces are rough, and the roughness has strong effects on many interesting problems, like wetting, adhesion, friction, light scattering, and surface growth, that are at the forefront of science and technology. The noise caused by the irregular fluctuations on a rough surface is a fundamental source of disorder. It plays an important role in the investigation of the physical properties of rough surfaces and interfaces. The quenched noise, which does not change with time, is usually more important than that of temporal noise [1–3] and will be the main focus in this chapter. The use of fractal concepts in understanding rough surface is increasingly important these days. The self-affine roughness exponent ($\alpha < 1$) not only defines the scaling properties of the surface, but also plays an essential role in describing the structure-property relationships of interfaces.

Both the long-range noise correlation function and the height correlation function are going to be derived as the result of the interactive fluctuations of an irregular surface, where distant events may have influenced each other. These two functions are needed in the study of the effects of roughness on the wetting and adhesion, the contact line depinning, and the critical surface tension. Beyond statics, we shall investigate the dynamics of wetting, the microscopic friction of surfaces in relative motion, and the diffusive light scattering. The surface adhesion and the bulk deformation are the two main contributions to friction that will be discussed. Finally, we are going to look at the surface growth and its relation to the microscopic roughness via the useful concept of dynamic scaling.

8.1 Fractal Surfaces

A rough surface is usually described by its deviation from a smooth reference surface. The statistical methods are adequate in explaining the spread of heights above or below this reference and the variation of these heights along the surface. Fractal, however, presents a nature language for interpreting the scaling behavior of roughness on all length scales. Self-similar fractals are isotropic and invariant under isotropic dilatation (see Section 4.10 and Appendix 4A). In contrast, surfaces are invariant under anisotropic transformations and belong to the broad class of self-affine fractal [1–4]. Both statistical and fractal techniques will be used in the following to describe the microstructure of rough surfaces.

The height of a continuous rough surface from its smooth reference is represented by the function $h(\vec{r})$, where $\vec{r}$ is the position vector on the reference surface. The distribution of surface heights is given by the height distribution function $\rho(h)$. It is normal to ensure that

$$\langle h \rangle = \int_{-\infty}^{\infty} h\rho(h)\,dh = 0. \tag{8.1}$$

The root mean square height of the surface is equal to the standard deviation (σ) and given by

$$\sigma = \left[\int_{-\infty}^{\infty} h^2 \rho(h)\,dh \right]^{1/2} \equiv [\langle h^2 \rangle]^{1/2}. \tag{8.2}$$

In general, the height distribution does not have to be Gaussian. In addition to the standard deviation that characterizes the fluctuation normal to the surface in the z-direction, it is reasonable to expect a correlation length ξ parallel to the surface. The correlation length can be defined by a correlation of fluctuations of $h(\vec{r})$ at two points $\vec{r}_o$ and $\vec{r}_o + \vec{r}$

$$\psi(\vec{r}) \equiv \langle h(\vec{r}_o)h(\vec{r}_o + \vec{r}) \rangle - \langle h(\vec{r}_o) \rangle^2, \tag{8.3}$$

which goes to zero when the two heights become uncorrelated at the distance of the order of ξ. Therefore,

$$\xi = \int r\psi \, dr \Big/ \int \psi \, dr. \tag{8.4}$$

Instead of ψ, very often the height correlation function

$$C(\vec{r}) \equiv \langle [h(\vec{r}_o + \vec{r}) - h(\vec{r}_o)]^2 \rangle = \sigma^2 - \psi(\vec{r}) \tag{8.5}$$

is considered.

Fractal is a useful concept for surfaces that are rough on all length scales. For a self-affine surface, an exponent α exists smaller than one such that the transformation,

$$\vec{r} \to m\vec{r}, \qquad z \to m^\alpha z \tag{8.6}$$

leaves the surface statistically invariant with the scaling factor m assumed to be $0 < m < 1$. The exponent α is a measure of the surface roughness. The change of the height correlation function with distance r is given by

$$C(\vec{r}) = \langle [\Delta h(r)]^2 \rangle \sim r^{2\alpha}, \qquad r \ll \xi, \tag{8.7}$$

where $\Delta h(\vec{r}) = h(\vec{r}_o + \vec{r}) - h(\vec{r}_o)$ from Eq. (8.5). At long range, the global behavior is described by $C(\vec{r}) = \sigma^2$ for $r \gg \xi$. The roughness exponent α is related to the spatial dimension (d) and the local fractal dimension (d_f) by $\alpha = d - d_f$. The roughness exponent defines the scaling properties of the surface. In the case of Gaussian distribution

$$\rho(h) = \frac{1}{\sqrt{2\pi r}} \exp\left(-\frac{h^2}{2r}\right), \tag{8.8}$$

eqs. (8.7) and (8.8) give $C \sim r$, which means $\alpha = 1/2$ and $d_f = 3/2$ for the two-dimensional surface $(d = 2)$.

The local slope follows directly from Eq. (8.7):

$$\frac{d|\Delta h|}{dr} \sim \frac{1}{r^{1-\alpha}}, \tag{8.9}$$

where $|\Delta h| = \langle [\Delta h(r)]^2 \rangle^{1/2}$. This slope decreases as the size of chosen domain (r) increases. For a given domain size, the smaller α corresponds to rougher local variations of a surface, and smoother hills and valleys are expected as $\alpha \to 1$. This physical picture is supported by measurements of the short-range surface profiles [5]. For non-Gaussian surfaces, three independent parameters (σ, ξ, α) are needed to describe the microstructure of a rough surface as shown in Figure 8.1.

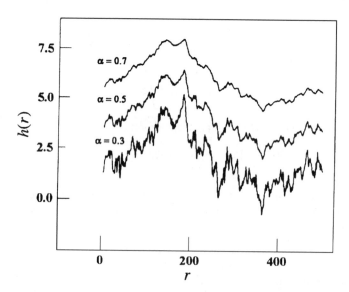

FIGURE 8.1. Surface profiles at different values of the roughness exponent (α). The standard deviation (σ) and correlation length (ξ) are assumed to remain the same. The scales are in arbitrary units [5].

8.2   Noise and Fluctuations

The irregular fluctuations of a rough surface are analyzed as stochastic processes expressed by Brownian motion. We would like (1) to seek a better understanding of the long-range noise correlation and (2) to derive an explicit expression that links the noise correlation function $\langle \mu(0)\mu(\vec{r})\rangle$ to the microstructure (σ, ξ, α) of rough surfaces in the discussion of long tails [6]. In a disordered medium, the quenched noise is usually more important than that of temporal noise [1]. Therefore, we shall focus on the quenched disorder, which does not change with time and is much more important than the thermal noise that is always present.

An exact solution of the height correlation function in the entire range of r can be determined by the following stochastic differential equation, which describes the fluctuations of local slope on a rough surface

$$\frac{d(\Delta h)}{dr} = -\frac{\Delta h}{2\xi} + \mu(r). \tag{8.10}$$

The first term on the right-hand side is the average local slope, and μ is the noise term that is the source of fluctuations of the local slope. The nature of noise is described by a correlation function. Our main purpose is to derive a nontrivial average of the quenched noise along a given interface. Eq. (8.10) has the form of the Langevin equation for Brownian motion (see Section 2.2), whose concepts and methods are applicable to a wide class of physical phenomena. Here, the velocity of a Brownian particle is replaced by Δh, the variable time by r, and the frictional coefficient by $(2\xi)^{-1}$. Integrating Eq. (8.10), squaring it, and taking the mean, we get

$$\langle [\Delta h(r)]^2 \rangle = \exp(-r/\xi) \int_0^r \int_0^r \exp[(r_1 + r_2)/2\xi] \langle \mu(r_1)\mu(r_2)\rangle \, dr_1 \, dr_2. \tag{8.11}$$

Our main interest is the noise correlation function. In cases in which the statistics of noise is Gaussian, as is assumed in most applications, one has the uncorrelated white noise with $\langle \mu \rangle = 0$ and

$$G(r_1 - r_2) \equiv \langle \mu(r_1)\mu(r_2)\rangle = \chi \delta(r_1 - r_2). \tag{8.12}$$

The constant χ is determined by the requirement of $\langle \Delta h^2 \rangle = \sigma^2$, which gives $\chi = \sigma^2/\xi^2$. Substituting Eq. (8.12) into Eq. (8.11) yields

$$C(r) = \sigma^2[1 - \exp(-r/\xi)] = \sigma r/\xi + \cdots, \quad \text{for } r \ll \xi. \tag{8.13}$$

Comparing eqs. (8.7) and (8.13), we obtain (1) $\alpha = 1/2$ and (2) a generalization of Eq. (8.13) to the cases of $\alpha \neq 1/2$ as

$$C(r) = \sigma^2\{1 - \exp[-(r/\xi)^{2\alpha}]\}. \tag{8.14}$$

A straightforward mathematical deduction is used in the derivation of Eq. (8.14). We have exponentiated an infinite series similar to that of Eq. (8.13). The infinite

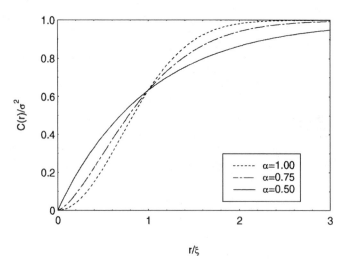

FIGURE 8.2. The height correlation functions at different values of the roughness exponent (α) that defines the short-range surface profile.

series, $1 - (r/\xi)^{2\alpha} + \cdots$, is uniformly convergent in $0 \le r/\xi \le 1$. At the same time, the roughness exponent α happens to be significant mainly within the domain of convergence, as shown in Figure 8.2. Eq. (8.14) is in agreement the numerical result obtained from computer simulation [1].

Spectral distribution is often used in discussing the observed surface topography [7]. An important measure of the surface statistics is the autocorrelation function $\psi(r)$ that is real. For stationary surfaces, the autocorrelation function is expressed by the power spectral density $\overline{\psi}(q)$ by a Fourier transform:

$$\overline{\psi}(q) = \int_{-\infty}^{\infty} \psi(r)\exp(-iqr)\,dr = 2\,\mathrm{Re}\int_{0}^{\infty} \psi(r)\exp(-iqr)\,dr, \quad (8.15)$$

where q is the spatial frequency of the undulations on the surface. The function $\psi(r)$ is given by the second term in Eq. (8.14) in accordance with Eq. (8.5). Two special cases can be calculated analytically:

$$\overline{\psi}(q) = \frac{2\sigma^2\xi}{1 + (q\xi)^2}, \quad \text{for } \alpha = 1/2 \quad (8.16)$$

and

$$\overline{\psi}(q) = \sqrt{\pi}\sigma^2\xi \cdot \exp[-(q\xi)^2/4], \quad \text{for } \alpha = 1. \quad (8.17)$$

These two functions are illustrated in Figure 8.3, which provides us with the essential information about the noise and fluctuations of rough surfaces. A broadband of continuous power spectrum [8] marks chaotic behavior. In Figure 8.3, we see such broad spectrum and the relatively uninformative region $q\xi < 1$. Therefore, we

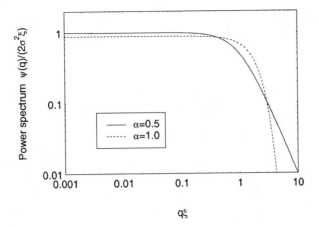

FIGURE 8.3. The surface spectral power $\overline{\psi}(q)$ versus the non-dimensional frequency $q\xi$ of rough surfaces.

shall put our attention to the region $q\xi > 1$ in the following study of the correlated noise.

Each Langevin equation has a corresponding Fokker–Planck equation. From this, we can have an integral equation that establishes the relation between the autocorrelation function $\psi(r)$ and the noise correlation function $G(r)$ [see Eq. (2.58)]:

$$\frac{d\psi(r)}{dr} = -\frac{1}{\sigma^2} \int_0^r G(r-s)\psi(s)\,ds. \qquad (8.18)$$

This integral equation follows directly from a statistical average of the Fokker–Planck equation for Brownian motion, and it has automatically taken into account the non-Gaussian memory effect. Replacing $iq = p/\xi$ in Eq. (8.15), we consider the Laplace transform of Eq. (8.18)

$$\overline{G}(p) = \int_0^\infty G(u)\exp(-pu)\,du = -\chi\,\frac{p\overline{\psi}(p) - \sigma^2}{\overline{\psi}(p)}, \quad \text{with } \chi = \sigma^2/\xi^2. \qquad (8.19)$$

According to a mathematical theorem:

$$\lim_{p\to\infty} p\overline{\psi}(p) = \lim_{u\to 0} \psi(u), \qquad (8.20)$$

we write the series expansion of the autocorrelation function:

$$\psi(u)/\sigma^2 = \exp(-u^{2\alpha}) = 1 - u^{2\alpha} + \frac{u^{4\alpha}}{2!} - \frac{u^{6\alpha}}{3!} + \cdots, \quad \text{with } u = r/\xi. \qquad (8.21)$$

Eqs. (8.18) and (8.21) clearly reveal the non-Gaussian characteristics of a non-Markovian process. Taking the Laplace transform of Eq. (8.21) and then substituting it into Eq. (8.19), we obtain

$$\overline{G}(p) = -\chi \frac{p \sum_{n=1}^{\infty} (-1)^n \frac{\Gamma(2\alpha n+1)}{n! p^{2\alpha n}}}{1 + \sum_{n=1}^{\infty} (-1)^n \frac{\Gamma(2\alpha n+1)}{n! p^{2\alpha n}}}, \tag{8.22}$$

where Γ is the gamma function. When $\alpha = 1/2$, we get $\overline{G}(p)/\chi = 1$, whose Laplace inversion is the delta function mentioned in Eq. (8.12). The Laplace inversion of the leading term in Eq. (8.22) gives

$$\langle \mu(0)\mu(r) \rangle = \frac{\sigma^2}{\xi^2} \cdot \frac{\Gamma(2\alpha + 1)}{\Gamma(2\alpha - 1)} \cdot \left(\frac{r}{\xi}\right)^{2\alpha-2}, \quad \text{for } 1/2 < \alpha < 1 \tag{8.23}$$

in the limit of $r/\xi \ll 1$. Of course, Eq. (8.22) is needed for $0 \le r/\xi < 1$. Eq. (8.23) is the sought after equation for the long-range noise correlation in space that is ubiquitous in nature. Interestingly, the long tail is directly related to the roughness exponent and correlation length. The strength of the noise correlation depends also on the standard deviation σ. Eq. (8.23) is good for describing the anomalous behavior of rough interfaces observed experimentally with the roughness exponent $\alpha \approx 0.6 - 0.8$ [3] that exceeds $\alpha = 1/2$. Furthermore, this model will play a prominent role in many high-technology applications involving wetting, adhesion, and diffuse scattering.

8.3   Fluctuations of Contact Line

When a surface is not perfectly smooth, the shape of the surface under thermodynamic equilibrium can be obtained by the well-known formulae in differential geometry. They are somewhat complicated. In general, however, they are considerably simplified when the surface deviates only slightly from a plane $z = 0$. Under such assumption, the area A of the surface is given by the integral [9]

$$A = \int \sqrt{1 + \left(\frac{\partial h}{\partial x}\right)^2 + \left(\frac{\partial h}{\partial y}\right)^2}\, dx dy. \tag{8.24}$$

For small h, it is approximated by

$$A = \int \left[1 + \frac{1}{2}\left(\frac{\partial h}{\partial x}\right)^2 + \frac{1}{2}\left(\frac{\partial h}{\partial y}\right)^2\right] dx dy. \tag{8.25}$$

The variation δA is

$$\delta A = \int \left[\frac{\partial h}{\partial x}\frac{\partial \delta h}{\partial x} + \frac{\partial h}{\partial y}\frac{\partial \delta h}{\partial y}\right] dx dy.$$

Integrating by parts, we find

$$\delta A = -\int \left[\frac{\partial^2 h}{\partial x^2} + \frac{\partial^2 h}{\partial y^2}\right] \delta h\, dx dy.$$

Hence, the required condition for the surface of equilibrium thin film is the Laplace equation:

$$\frac{\partial^2 h}{\partial x^2} + \frac{\partial^2 h}{\partial y^2} = 0. \tag{8.26}$$

When three adjoining media are in equilibrium, the surfaces of separation are such that the resultant of the surface tension forces is zero on common line of intersection. This condition implies that the surfaces of separation must intersect at angles called contact angles that are determined by the values of surface tensions (see Section 8.4).

Let us construct the shape of interface $h(x, y)$ associated with the three-phase contact line. The contact line has the form $x = \lambda(y)$ that is covered by vapor in the domain of $-\infty < x < \lambda(y)$ and by liquid in the domain of $\lambda(y) < x < \infty$, as shown in Figure 8.4. The solution of Eq. (8.26) is

$$h(x, y) = \theta_o x + \frac{1}{2\pi} \int_{-\infty}^{\infty} \tilde{h}(q) \exp(iqy - |q|x) \, dq. \tag{8.27}$$

The first term is the unperturbed profile on a smooth surface, and θ_o is the equilibrium angle of contact. The second term describes the correction because of the roughness of wavelength $2\pi/|q|$. Imposing that h vanishes on the contact line, $h[\lambda(y), y] = 0$, we are able to relate the amplitude $\tilde{h}(q)$ to the Fourier component

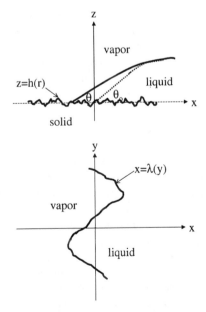

FIGURE 8.4. Definitions of the contact angle and the contact line.

of the line shape,

$$\tilde{h}(q) = -\theta_o \tilde{\lambda}(q) = -\theta_o \int_{-\infty}^{\infty} \lambda(y) \exp(-iqy) \, dy. \tag{8.28}$$

Substituting this equation into Eq. (8.27) and inverting the Fourier transform, we obtain

$$h(x, y) = \theta_o(x - 1/\pi) + \int_{-\infty}^{\infty} \frac{x\lambda(y')}{x^2 + (y - y')^2} \, dy'. \tag{8.29}$$

This equation gives an explicit expression between the shape of the interface and the profile of the contact line.

The free energies associated with the contact line are important to the later study of wetting and adhesion. For inhomogeneous surfaces, the correction to the capillary energy is

$$\Delta U_{cap} = \frac{1}{2} \int \gamma \left[(\nabla h)^2 - \theta_o^2 \right] dx \, dy, \tag{8.30}$$

which follows from Eq. (8.25), where γ is the interfacial free energy between the liquid and vapor interface. Integrating over the entire region of $\lambda(y) < x < \infty$, we obtain

$$\Delta U_{cap} = \frac{\gamma \theta_o^2}{2} \int_{-\infty}^{\infty} |q| |\Lambda(q)|^2 \frac{dq}{2\pi}. \tag{8.31}$$

This energy develops from the increase in the surface area of the liquid–vapor interface. The unusual $|q|$ dependence of the energy function comes from the integration of a q^2 energy over a distance $|q|^{-1}$ as a result of the contact line distortion. Eq. (8.31) was originally derived by Joanny and de Gennes [10] in their study of the deformation of a contact line with a small angle of contact.

8.4   Wetting and Adhesion

The wetting phenomena have been widely studied both theoretically [11] and experimentally [12] in connection with the physics of surfaces and interfaces. The behavior of liquid partially wetting a smooth solid surface is well understood. The case of rough solid surfaces, however, is much less clear, even though roughness is a real-world problem and its value in practical applications is high. Therefore, the attention here is to the understanding of the effect of the surface microstructure on the partial wetting phenomena. We shall analyze the change in contact angle and determine how it is coupled to the wandering of three-phase contact line because of the microstructure disorder of a rough surface. It is assumed that the wetting fluid spreads slowly and the macroscopic wetting phenomena will be linked to the noise correlation function beyond the white-noise limit.

To begin with an ideal situation of flat and smooth solid surfaces, the equilibrium angle of contact θ_o is determined by the Young equation:

$$\gamma_s - \gamma_{sl} = \gamma \cos \theta_o. \tag{8.32}$$

Here, γ_s, γ_{sl}, and γ are the interfacial free energy per unit area for the solid–vapor, solid–liquid, and liquid–vapor interfaces. The work of adhesion is defined by Young's equation as

$$w_{ad} = \gamma_s + \gamma - \gamma_{sl}. \tag{8.33}$$

Combining the above two equations yields the Dupre equation:

$$w_{ad} = \gamma(1 + \cos \theta_o). \tag{8.34}$$

By analyzing the molecular attractive forces interacting at the interface, the work of adhesion and the equilibrium contact angle can be linked directly to the Hamaker constant and the Lifshitz–van der Waals constant $\hbar\bar{\omega}$ [see Appendix 8A]. At the molecular level, the Planck constant $\hbar$ appears not only in the equilibrium theory of interfaces, but also in a nonequilibrium theory (see Section 8.6).

The contact line of a liquid partially wetting the smooth solid is a straight line chosen to be in the y-direction: $x = 0$. In the real situation of rough substrates, the Young–Dupre equation has to be generalized to include the spatial-dependent (x, y) interfacial energy densities and contact angle in the description of the local wetting phenomena,

$$\gamma_s(x, y) - \gamma_{sl}(x, y) = \gamma \cos[\theta(x, y)] \equiv \gamma \cos[\theta_o - \phi(x, y)], \tag{8.35}$$

where ϕ is caused by roughness. The contact free energy of the system is the sum of interaction energies with the substrate. From Figure 8.4, the contact free energy is written as

$$U_{con} = \int dy \int_{-\infty}^{\lambda(y)} \gamma_s(x, y)\, dx + \int dy \int_{\lambda(y)}^{\infty} \gamma_{sl}(x, y)\, dx$$

$$= \int dy \int_{-\infty}^{\lambda(y)} [\gamma_s(x, y) - \gamma_{sl}(x, y)]\, dx. \tag{8.36}$$

The integrals over y are taken over the entire system. The difference in energy between the rough and smooth surfaces is

$$\Delta U_{con} = U_{con} - \langle U_{con}\rangle = \int dy \int_{-\infty}^{\lambda(y)} \Delta w(x, y)\, dx, \tag{8.37}$$

where

$$\Delta w(x, y) = [\gamma_s(x, y) - \gamma_{sl}(x, y)] - \langle \gamma_s - \gamma_{sl}\rangle \tag{8.38}$$

is the local energy density. From eqs. (8.35) and (8.38), a spatial-dependent Wenzel roughness [13,14] can be written as

$$\varepsilon(x, y) = \frac{\Delta w(x, y)}{\langle \gamma_s - \gamma_{sl} \rangle} + 1 = \frac{\cos[\theta_o - \phi(x, y)]}{\cos \theta_o} \cong 1 + \theta_o \phi(x, y) + \cdots,$$

$$\text{for } \theta_o \ll 1. \quad (8.39)$$

This equation shows that $\Delta w(x, y)$ is proportional to the local slope $\phi(x, y)$ [11] because of surface roughness. As we shall see later that the Wenzel roughness and energy density are needed in the study of the critical surface tension and contact line depinning, respectively.

Roughness results in local changes in the contact angle and hence the shape of the contact line. Instead of being a straight line in the case of planar substrates, the three-phase contact line tends to wander on the x-y plane because of the roughness, as shown in Figure 8.4. The spatial-dependent angle of contact and locus of wedge intersection fluctuate on an irregular surface to find their optimal angle and position via the minimization of the total free energy of surface. Let us introduce the Fourier component of the local energy density,

$$\tilde{w}(q) = \int_{-\infty}^{\infty} w(y) \exp(-iqy) \, dy.$$

The capillary energy associated with the contact line is given by Eq. (8.31) as a result of the contact line distortion. The total change in the free energy is

$$\Delta U = \Delta U_{cap} + \Delta U_{con}$$

$$= \frac{\gamma \theta_o^2}{2} \int_{-\infty}^{\infty} |q| |\tilde{\lambda}(q)|^2 \frac{dq}{2\pi} + \int_{-\infty}^{\infty} \Delta \tilde{w}(q) \tilde{\lambda}(q) \frac{dq}{2\pi}. \quad (8.40)$$

Minimizing ΔU with respect to $\tilde{\lambda}$, i.e., $d(\Delta U)/d\tilde{\lambda} = 0$, we obtain

$$\Delta W(q) = \gamma \theta_o^2 |q| \tilde{\lambda}(q). \quad (8.41)$$

We shall return to Eq. (8.41) when the energy density is determined.

Let us introduce a new concept of determining the effect of surface roughness on the wetting properties of surfaces from the noise correlation function. In terms of ϕ, given in Eq. (8.35), its autocorrelation function is related to the noise correlation function by $2\langle \phi(0)\phi(r) \rangle = \langle \mu(0)\mu(r) \rangle$. Taking the spatial average over the range of the correlation length ξ, the noised-induced wetting is determined by [15]

$$\overline{\phi} = \left[\frac{1}{2\xi} \int_0^\xi \langle \mu(0)\mu(r) \rangle dr \right]^{1/2}. \quad (8.42)$$

This equation clearly follows the same concept as that for the root mean square height mentioned earlier. In the case of uncorrelated white noise, the noise correlation function is given by Eq. (8.12), and Eq. (8.42) gives $\overline{\phi} = \sigma/2\xi$, which provides a consistent check between eqs. (8.10), (8.12), and (8.42).

In the case of noise that has long-range correlation, Eq. (8.42) is still valid for the determination of the change in the contact angle. Using Eq. (8.23) and taking the spatial averages, we obtain the effective Wenzel roughness from eqs. (8.23), (8.39), and (8.42):

$$\bar{\varepsilon} = 1 + \theta_o \bar{\phi} + \cdots, \qquad \text{with } \bar{\phi} = C_\alpha \sigma/\xi + \cdots. \qquad (8.43)$$

According to Wenzel, $\bar{\varepsilon}$ is the ratio of nonplanar to planar surface area and $\bar{\varepsilon} = 1$ for the planar system. Eq. (8.43) shows that roughening the surface reduces the contact angle and promotes wetting. The effective energy density is

$$\overline{\Delta w} = \gamma \theta_o \bar{\phi} = C_\alpha \gamma \theta_o \sigma/\xi + \cdots. \qquad (8.44)$$

By definition, the spatially independent $\overline{\Delta w}$ is actually equal to the difference in the work of adhesions [see Eq. (8.33)] between the liquid on a rough surface and that on a smooth surface. The nondimensional C_α is

$$C_\alpha = \left[\frac{1}{2(2\alpha - 1)} \frac{\Gamma(2\alpha + 1)}{\Gamma(2\alpha - 1)} \right]^{1/2}, \quad \text{for } 1/2 < \alpha < 1. \qquad (8.45)$$

Clearly, eqs. (8.23) and (8.43)–(8.45) reveal that the effect of the long-range noise correlation on $\bar{\varepsilon}$ and $\overline{\Delta w}$ is expressed by α having a value differing from $1/2$. C_α is a monotonic increasing function of the roughness exponent and is proportional to $\sqrt{\alpha}$. Of course, the general expression, Eq. (8.22), is needed for higher approximations. For the uncorrelated noise, we get $\alpha = 1/2$, $C_\alpha = 1/2$, and Eq. (8.12). In addition to the fluctuations (σ, ξ) of fractal surfaces and the equilibrium properties (γ, θ_o) of smooth substrates, our analysis reveals that the roughness-induced wetting increases not only with the roughness exponent, but also with the long-range noise correlation. The fractal structure beyond the white-noise assumption plays an important role.

We are now in the position of returning to Eq. (8.41). Taking the spatial average of the Fourier inversion of Eq. (8.41) and using Eq. (8.45) give

$$\overline{\left| \frac{d\lambda(y)}{dy} \right|} = \frac{1}{\theta_o} \bar{\phi} = \frac{C_\alpha \sigma}{\theta_o \xi} + \cdots, \qquad (8.46)$$

where the average is taken over the range of the correlation length ξ. This equation shows that the slope of the contact line increases with roughness as a function of its strength and long-range noise correlation. By definition, the spatial-independent slope is $\bar{\phi} = \overline{|dh(r)/dr|}$. Therefore, Eq. (8.46) suggests that the slope $\lambda'(y)$ amplifies that of a smooth surface by a factor of $1/\theta_o$. As a result, the three-phase contact line is broadened by the roughness of the surface.

8.5   Critical Surface Tension

The surface energy (or tension) of solids may not be measured directly because of the viscoelastic constraint of the bulk phase, which necessitates the use of indirect methods. Zisman use extrapolated contact angle measurements to define the critical surface tension (γ_{co}) for planar substrates as [12]

$$\cos\theta_o = 1 + b(\gamma_{co} - \gamma), \qquad (8.47)$$

where $b = -d(\cos\theta_o)/d\gamma > 0$ is Zisman's slope of a smooth substrate. γ_{co} is characterized as an important property of solid surface.

In the case of rough surfaces, we can write the effective critical surface tension ($\overline{\gamma}_c$) in the same form,

$$\cos(\theta_o - \overline{\phi}) = 1 + b'(\overline{\gamma}_c - \gamma), \qquad (8.48)$$

where $b' = -d(\cos\theta_o - \overline{\phi})/d\gamma > 0$ is the yet-to-be-determined Zisman's slope of a rough substrate. Eqs. (8.47) and (8.48) are related to the effective Wenzel roughness by

$$\cos(\theta_o - \overline{\phi}) = \overline{\varepsilon}\cos\theta_o. \qquad (8.49)$$

Combining eqs. (8.47)–(8.49) and using the condition of $\overline{\gamma}_c = \gamma_{co}$ at $\overline{\phi} = 0$, we get

$$\overline{\gamma}_c = \gamma_{co} + \frac{\overline{\varepsilon} - 1}{b\overline{\varepsilon}} \qquad (8.50)$$

and

$$b' = \overline{\varepsilon}b = -\overline{\varepsilon}\frac{d(\cos\theta_o)}{d\gamma} > 0. \qquad (8.51)$$

Consider $\gamma_{co} = 20$ dyne/cm and $b = 0.025$ for a polymeric surface. In Figure 8.5, the influence of surface roughness on the Zisman plot is calculated from

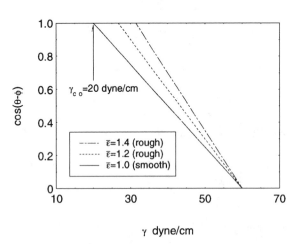

FIGURE 8.5. Effect of surface roughness on the Zisman plot. $\overline{\varepsilon}$ is the effective Wenzel roughness.

eqs. (8.48) to (8.51). Higher critical surface tension and steeper slope are seen for rougher surfaces. Figure 8.5 also shows that $\overline{\phi} > 0$ for $\theta_o < 90°$. Substituting Eq. (8.43) into Eq. (8.50) yields

$$\overline{\gamma}_c = \gamma_{co} + \frac{\theta_o C_\alpha \sigma}{b\xi} + \cdots . \tag{8.52}$$

As the roughness increases, the critical surface tension of the solid is raised from γ_{co} to a value that depends on the strength and range of the correlated noise. The wetting transition from the partial wetting to the complete wetting is determined by the critical surface tension. For practical applications, it becomes evident from Eq. (8.52) that the physical and chemical treatment of low-energy surfaces to improve the adhesion may change $\overline{\gamma}_c$ because of a change in surface chemistry or simply increase $\overline{\gamma}_c$ because of surface roughness.

To recapture the main points of the last three sections, a useful concept has been developed that enables us to determine the effect of surface roughness on the wetting properties of surfaces from the noise correlation function. The long-range correlated noise is caused by the interactive fluctuations of an irregular surface and is treated as the source of disorder in the determination of the noise-induced wetting. The present study goes beyond the familiar white-noise assumption. Analytical expressions and functional relationships of the contact angle, the critical surface tension, and the contact line depinning have been derived as a function of the microstructure of a rough surface. In addition to the fluctuations (σ, ξ) of fractal surfaces and the equilibrium properties (γ, θ_o) of smooth substrates, the noise-induced wetting depends not only on the roughness exponent (α), but also on the strength and range of the correlated noise. Roughening the surface reduces the contact angle, increases the critical surface tension, and broadens the three-phase contact line.

8.6 Dynamics of Wetting

The formation of an interface is controlled not only by the statics, but also by the dynamics of wetting. This process is particularly important to the technology application for which it may not be possible to allow the necessary time for a system to reach its equilibrium. The Young–Dupre equation for the smooth solid surface has to be extended to include both time and roughness dependence into its formula. What the nonequilibrium states of a system are depends on its rates of relaxation and the time allowed for the system to reach the equilibrium. Therefore, we would like to establish the link between the distribution of the relaxation times and the variation of the microscopic length scales on a rough interface.

Let us again start with the situation of flat surfaces. The dynamics of wetting at the molecular level has been treated by adapting Eyring's rate theory [16]. The viscosity of the liquid controls the forward and backward movements of the nonequilibrium contact line between liquid and substrate in this rate process, and the dynamics of wetting has been adequately described by a single relaxation time for smooth substrates [17]. In the case of rough substrates, a distribution of the relaxation times is needed to take into account all of the local length scales on a rough interface. By considering a discrete description of the distribution in a nonequilibrium system, the difference in the interfacial-energy densities at a given locale denoted by the subscript "i" is related to the corresponding difference in the contact angles by

$$\Delta U_i(t) = [\gamma_s(t) - \gamma_{sl}(t)]_i - [\gamma_s^{(\infty)} - \gamma_{sl}^{(\infty)}]_i$$

$$= \gamma[\cos\theta_i(t) - \cos\theta_i^{(\infty)}] \equiv \gamma \cdot \Delta\cos\theta_i(t) \qquad (8.53)$$

[see Eqs. (8.32) and (8.38)]. Thus, the total energy difference of the system is

$$\Delta U(t) = \sum_i \gamma\Delta\cos\theta_i(t) \equiv \gamma\Delta\cos\theta(t), \qquad (8.54)$$

which also defines the effective dynamic angle of contact $\theta(t)$.

The influence of the surface microstructure on the nonequilibrium wetting will be analyzed as a random stochastic process coupled with the above-mentioned original concept of Eyring. The advancing and retreating of the three-phase contact line is no longer satisfied by the detailed balance but by the Master equation [Eq. (3.88)] that is pivotal to the time-dependent behavior, from which the simplest governing equation for the dynamic angle of contact at the ith locale can be written as

$$\frac{d\Delta\cos\theta_i(t)}{dt} = \Delta\cos\theta_{if}\Lambda_{if} - \Delta\cos\theta_{ib}\Lambda_{ib}. \qquad (8.55)$$

Here, the subscript "f" and "b" refer the "forward" and "backward" movements of the contact line, respectively. When the activation energy for the viscous flow of the liquid is E_η, the change in free energy caused by the action of surface forces ΔW_i affects the transition probabilities per unit time [18],

$$\Lambda_{if} = \frac{kT}{2\pi\hbar}\exp\left(-\frac{E_\eta + \Delta W_i/2}{kT}\right) \qquad (8.56)$$

and

$$\Lambda_{ib} = \frac{kT}{2\pi\hbar}\exp\left(-\frac{E_\eta - \Delta W_i/2}{kT}\right), \qquad (8.57)$$

where k is Boltzmann's constant and $\hbar$ is Planck's constant. We have $\Delta W_i = \gamma \cdot (area)_i = \gamma\Omega/L_i$, where Ω is the activation volume and L_i is the ith local length scale. When the angles of contact are small, we have $\theta_{if} = \theta_{ib} = \theta_i(t)$, which imply $\Delta W_i \ll 2kT$. From eqs. (8.56) and (8.57), the ith relaxation time can be written as

$$\tau_i = (\Lambda_{if} - \Lambda_{ib})^{-1} = \frac{L_i}{\gamma} \cdot \frac{2\pi\hbar}{\Omega}\exp\left(\frac{E_\eta}{kT}\right) = \frac{\eta L_i}{\gamma}. \qquad (8.58)$$

Hence, Eq. (8.55) becomes

$$\frac{d\cos\theta_i(t)}{dt} = -\frac{\cos\theta_i(t) - \cos\theta_i^{(\infty)}}{\tau_i}, \qquad i = 1, 2, \ldots, M, \qquad (8.59)$$

where the superscript (∞) represents equilibrium. This equation has been experimentally verified in the case of smooth interface [17], which has $M = 1$. The solution of Eq. (8.59) is

$$\cos\theta_i(t) = \cos\theta_i^{(0)} + \left[\cos\theta_i^{(\infty)} - \cos\theta_i^{(0)}\right] \cdot [1 - \exp(-t/\tau_i)], \qquad (8.60)$$

where $\cos\theta_i^{(0)} \equiv \cos\theta_i(t = 0)$ with $i = 1, 2, \ldots, M$. Eq. (8.54) and Eq. (8.60) yield

$$\frac{\cos\theta(t) - \cos\theta^{(0)}}{\cos\theta^{(\infty)} - \cos\theta^{(0)}} = 1 - \sum_i g_i \exp(-t/\tau_i) \equiv 1 - \Phi(t). \qquad (8.61)$$

Here,

$$g_i = \frac{\cos\theta_i^{(\infty)} - \cos\theta_i^{(0)}}{\cos\theta^{(\infty)} - \cos\theta^{(0)}} = \frac{\Delta U_i^{(\infty)}}{\Delta U^{(\infty)}}, \qquad (8.62)$$

with $\sum_i g_i = 1$. It can be interpreted as the statistical distribution of the differences in the interfacial-energy densities induced by the surface roughness and g_i is an equilibrium property.

The relaxation function Φ in Eq. (8.61) is a summation of a distribution of an exponential, so that naturally a stretched exponential is present [see sections 3.2 and 4.6 and eqs. (5.40) and (6.12)],

$$\Phi(t/\tau) = \sum_{i=1}^{M} g_i \exp(-t/\tau_i) = \exp[-(t/\tau)^{\alpha}], \qquad (8.63)$$

where $\tau = \eta L/\gamma$ is the macroscopic relaxation and L is the characteristic length of the system. Both γ and η are not affected by roughness. By letting

$$t/\tau \rightarrow u \quad \text{and} \quad \tau_i/\tau = L_i/L \rightarrow s, \qquad (8.64)$$

the summation in Eq. (8.63) can be written as an integral

$$\exp(-u^{\alpha}) = \int_0^{\infty} g(s)\exp(-u/s)\,ds = \int_{-\infty}^{\infty} H(s)\exp(-u/s)d\ln s. \qquad (8.65)$$

Interestingly, the stretched exponential shown in this equation is consistent with the experimental obsevation [19] as a logical consequence of the experimental geometry of rough surfaces. The Laplace inversion of Eq. (8.65) gives

$$H(s) = g(s)/s = \frac{1}{\pi s}\int_0^{\infty}\exp\left[-\left(\frac{u}{s} + u^{\alpha}\cos\pi\alpha\right)\right]\sin(u^{\alpha}\sin\pi\alpha)\,du. \qquad (8.66)$$

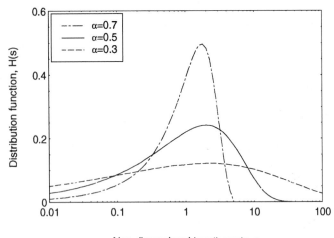

FIGURE 8.6. The distribution function $H(s) = g(s)/s$ is plotted against the non-dimensional length scale $(L_i/L \rightarrow s)$ at different values of the roughness exponent α.

In the case of $\alpha = 1/2$, Eq. (8.66) reduces to

$$H(s) = \frac{1}{\pi s} \int_0^\infty \exp(-u/s) \sin(\sqrt{u})\, du = \frac{s^{1/2}}{2\sqrt{\pi}} \exp\left(-\frac{s}{4}\right). \qquad (8.67)$$

This equation is a result of the Gaussian distribution [see Eq. (4.104)].

When the interface is smooth, we have $\alpha = 1$ and the distribution $g(s)$ is a delta function. The effect of the exponent α on the roughness-induced length spectrum $H(s)$ is calculated in Figure 8.6 on the basis of Eq. (8.66). $H(s)d\ln s$ gives the contribution to the wetting dynamics associated with the length scales between $\ln(s)$ and $\ln(s) + d\ln(s)$. This distribution becomes narrower and the interface becomes smoother as α increases. The physical picture obtained in Figure 8.6 is exactly the same one following Eq. (8.9), in which the local slope of the rough surface is scaled by the exponent α and decreases as the size of chosen domain $(r \sim L \ll \xi)$ increases. At this point, it becomes apparent that the α introduced in Eq. (6.63) is the *roughness exponent* presented in sections 8.1 and 8.2 that defines the scaling properties of self-affine fractal surfaces. In the special cases of $\alpha = 1$ and $1/2$, we have obtained the exact verifications of α as the roughness exponent.

From eqs. (8.61) and (8.63), a more explicit expression for the effective dynamic angle of contact is

$$\cos\theta(t) = \cos\theta^{(0)} + [\cos\theta^{(\infty)} - \cos\theta^{(0)}] \cdot \{1 - \exp[-(t/\tau)^\alpha]\}, \quad 0 < \alpha \leq 1. \qquad (8.68)$$

Here, $\theta^{(0)}$ is the initial angle of contact, and $\theta^{(\infty)} \equiv \theta(t \rightarrow \infty) = \theta_o - \overline{\phi}$ is the equilibrium angle of contact. Both θ_o and $\overline{\phi}$ are mentioned in eqs. (8.32) and (8.42), respectively. The effective relaxation time (τ) is a composite of the characteristic

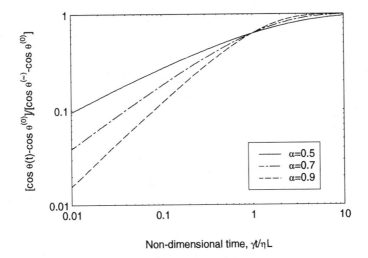

FIGURE 8.7. The calculated master curves of the effective dynamic angle of contact, $\cos\theta(t)$, as a function of the non-dimensional time and roughness exponent.

length, surface tension, and viscosity. The effect of roughness exponent on the dynamics of wetting is calculated in Figure 8.7. The roughness exponent has a much stronger effect on wetting when the system is far from equilibrium with smaller values of $\gamma t/\eta L$. Equilibrium is approached when the nondimensional parameter $\gamma t/\eta L$ is much greater than ten that is consistent with experimental data [17]. Both γ and η are usually a function of temperature. For polymers, a typical value for the temperature coefficient of surface tension, $-(1/\gamma)(\partial\gamma/\partial T)$, is close to 0.002 °C^{-1}. On the other hand, the viscosity is a much stronger function of temperature [see eqs. (8.58) and (5.45)]. The increase in the parameter $\gamma t/\eta L$ with temperature is therefore expected to be as a result of the decrease in the melt viscosity rather than the decrease in surface tension. From Figure 8.7 and Eq. (8.34), it becomes clear that the joint strength expressed by the dynamic work of adhesion, $w_a(t) = \gamma[1 + \cos\theta(t)]$, increases with temperature of joint formation, which has been observed experimentally.

Next, we shall calculate the influence of nonequilibrium on the transition from the complete wetting to the partial wetting of rough surfaces by using Eq. (8.68). From Section 8.5, the effective equilibrium angle of contact is

$$\cos\theta^{(\infty)} = 1 - b\bar{\varepsilon}(\gamma - \gamma_c), \quad \text{for } \gamma \geq \gamma_c \tag{8.69}$$

and

$$\cos\theta^{(\infty)} = 1, \quad \text{for } \gamma \leq \gamma_c, \tag{8.70}$$

where $\bar{\varepsilon}$ is the effective Wenzel roughness and γ_c is the critical surface tension of rough surface given by Eq. (8.52). Choose $\gamma_{co} = 20$ dyne/cm, $b = 0.025$, and $\theta^{(0)} = \theta_o = 10^0$ as the input. The numerical examples are shown in Figure 8.8.

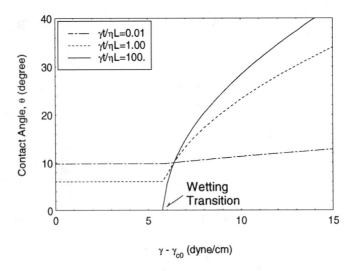

FIGURE 8.8. The influence of the non-dimensional time on the angle of contact and the wetting transitions in the vicinity of the critical surface tension. The microstructure parameters of the rough surface are $\alpha = 0.7$ and $\sigma/\xi = 1.0$.

Although the wetting transition can be clearly seen for a system close to equilibrium with $\gamma t/\eta L = 100$ (see Figure 8.8), the transition becomes less noticeable for systems far from equilibrium as $\gamma t/\eta L$ is getting smaller.

In this section, we have analyzed the dissipative and irreversible behavior of a liquid partially wetting a self-affine fractal surface at the molecular level and have derived the equations for predicting the dynamics of wetting as a function of three roughness parameters: σ, ξ, and α. The roughness exponent α is found to be equal to the stretched exponent that characterizes the relaxation function and defines the distribution of the relaxation times for the time-dependent angle of contact on rough surfaces. The exponent α is theoretically derived as a logical consequence of the experimental geometry [19]. The dynamic model reveals that (1) the nonequilibrium transition from exponential kinetics for planar surface to stretched-exponential kinetics for rough surface and (2) the equilibrium transition from partial wetting to complete wetting becoming less noticeable at nonequilibrium conditions. Because viscosity affects the rate of approach to an equilibrium configuration, its strong dependence on temperature may serve as a driving force that is useful in controlling the nonequilibrium behavior of wetting and adhesion of rough interfaces in many technological applications.

8.7   Adhesional Friction

Two main factors contribute to friction generated between unlubricated solid surfaces in relative motion. The first factor is the adhesion that occurs at the regions

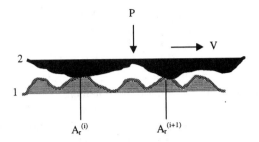

FIGURE 8.9. Contact between two rough surfaces.

of real contact and is usually more important. The second factor is caused by the bulk deformation, which will be discussed in the next section. Few real solids are smooth on the microscopic scale, and, consequently, the real area of contact (A_r) is confined to comparatively small portion of the apparent area of contact (A_a), where the asperities in the two surfaces meets (see Figure 8.9). At these points, the real interfacial contact is made. Motion of the surfaces over each other is possible only if the frictional energy starts to overcome the adhesion energy.

According to Eq. (8.33), the adhesion energy is $E_{ad} = \Delta\gamma \cdot A_r$, where $\Delta\gamma = \gamma_1 + \gamma_2 - \gamma_{12}$ is the work of adhesion for the two contacting surfaces referred to by the subscripts 1 and 2. When the contact is under the action of an applied normal load P, the required frictional energy for sliding to occur is $E_{fr} = fPd$, where d is the critical displacement at microscopic level in the direction of motion. By equating $E_{fr} = E_{ad}$, the coefficient of friction is defined as

$$f = \frac{\Delta\gamma}{d} \cdot \frac{A_r}{P}. \tag{8.71}$$

If both surfaces are of the same material, $\gamma_1 = \gamma_2 = \gamma$, $\gamma_{12} = 0$, and $\Delta\gamma = 2\gamma$. The ratio of $2\gamma/d$ is called the shear strength of the interface [20]. A typical value of the surface tension for a hydrocarbon or van der Waals surface is $\gamma = 20$ dyne/cm. Assuming $d = 1$ nm, we get the shear strength $2\gamma/d = 4 \times 10^8$ dyne/cm^2, which compares well with the experimental values of 2×10^8 dyne/cm^2 for surfaces sliding of mica in air. The variation of the friction coefficient with physical parameters, such as the surface energy, the roughness, and the applied load, resolves itself into the study of the effects of these factors on γ/d and A_r. We have mentioned in Section 8.5 that the surface tension of the solid γ is equal to the effective critical surface tension $\overline{\gamma}_c$ for systems in the absence of polar interactions. Therefore, the coefficient of friction is enhanced via the roughness-induced increase in the critical surface tension shown in eqs. (8.52).

If the deformation of the asperities is fully plastic (i.e., beyond the yield stress), the resistance to plastic deformation is measured by the hardness (H), which is the real pressure of the load over a real contact area. The real area of contact caused by one asperity is given by $A_r^{(i)} = P^{(i)}/H$, where $P^{(i)}$ is the load borne by that asperity. Hence, for an assembly of asperities that make the contacting area

$$A_r = \sum_i A_r^{(i)} = \sum_i P^{(i)}/H = P/H. \tag{8.72}$$

Eqs. (8.71) and (8.72) result in $f = \Delta\gamma/Hd$, which is independent of load. This model is usually good for metal but not for polymer.

It has been observed for polymers that the coefficient of friction may decrease with increasing load. Therefore, it appears that an explanation should be sought by the elastic deformation of the surfaces rather than the plastic deformation we have just discussed. The ratio of rough to planar surfaces is determined by the effective Wenzel roughness $\bar{\varepsilon}$ given by Eq. (8.43). When $\bar{\varepsilon}$ is close to one but is not equal to one, the real area of contact may be related to the apparent area of contact by

$$A_r \sim (\bar{\varepsilon} - 1)A_a. \tag{8.73}$$

If the deformation is Hertzian, we have [21]

$$A_a = (PR_\sigma/K)^{2/3}, \tag{8.74}$$

where

$$\frac{1}{K} = \frac{3}{4}\left(\frac{1 - v_1^2}{E_1} + \frac{1 - v_2^2}{E_2}\right) \tag{8.75}$$

and

$$R_\sigma^{-1} = R_{\sigma 1}^{-1} + R_{\sigma 2}^{-1}. \tag{8.76}$$

Here, E is the Young modulus, v is the Poisson ratio, and R_σ is the radius of curvature induced by roughness. Eqs. (8.71), (8.73), and (8.74) give $f \sim P^{-1/3}$, which is load dependent. In view of the fact that polymers deform viscoelastically, a more appropriate form of the friction coefficient for polymers would be [22]

$$f \sim P^{-(j-2)/j}, \quad \text{for } 2 < j < 3. \tag{8.77}$$

The condition of $j = 2$ corresponds the plastic contact, and $j = 3$ corresponds the elastic contact.

A quantitative description of the effect of roughness on friction remains unresolved. We would like to put together, however, what has already been analyzed for the elastic contact. For simplicity, one of the surfaces is now assumed to be rigid and flat. On the basis of the microstructure of fractal surfaces discussed in Section 2, the apparent radius of curvature of rough surfaces with $\alpha = 1$ is

$$R_\sigma = \xi^2/2\sigma, \tag{8.78}$$

which characterizes macroroughness. As one might expect, $R_\sigma \to \infty$ for smooth surface where the standard deviation $\sigma \to 0$ or the correlation length $\xi \to \infty$. Combining eqs. (8.43), (8.52), (8.71), (8.73), (8.74), and (8.78), we obtain

$$f \sim \frac{C_\alpha(\sigma\xi)^{1/3}K^{2/3}}{dP^{1/3}}\left(\gamma_{co} + \frac{\theta_o C_\alpha\sigma}{b\xi} + \cdots\right), \tag{8.79}$$

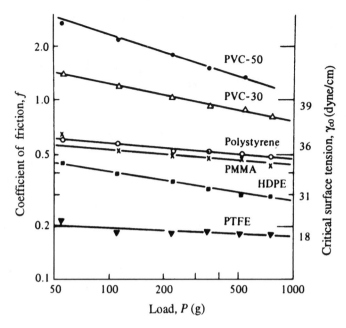

FIGURE 8.10. Correlation between the coefficient of adhesional friction, critical surface tension, and applied load. PTFE is polytetrafluoroethylene; HDPE is high-density polyethylene; PMMA is polymethyl methacrylate; PVC-30 and PVC-50 are polyvinyl chloride with two different levels of plasticiser [23].

where γ_{co} is the critical surface tension of smooth surface and C_α is given by Eq. (8.45). Eq. (8.79) reveals that smoother surfaces, which usually have larger ξ, may not result in a lower coefficient of friction unless σ is getting smaller at the same time. Therefore, lowering σ and ξ simultaneously may be needed for the purpose of reducing the coefficient of friction.

The effects of the critical surface tension and applied load on the friction coefficient described by Eq. (8.79) and Eq. (8.77) are qualitatively supported by the experimental data of a half dozen polymers [23], as shown in Figure 8.10. So far, we have been studying the case of dry sliding between rough surfaces in which the coefficient of adhesional friction usually dominates over the contribution from the bulk deformation. The use of a lubricant between rough surfaces in relative motion, however, virtually eliminates the adhesion term. The measured friction force is then attributed solely to the deformational component that will be analyzed in the next section.

8.8   Deformational Friction

The deformations that were brought about by the adhesion mechanism of friction in the last section were of the order of a nanometer. In the discussion that follows the deformation may extend over a larger volume exceeding a micron.

Consider that a rigid body slides smoothly over the surface of a polymeric solid. Because the material in front of the slider is compressed, work has to be done to overcome the materials resistance. Behind the slider, the viscoelastic recovery will assist the forward motion of the slider, but because of the viscoelastic losses associated with the material being subjected to a strain cycle, some of the energy input will be dissipated. The difference between the input work and the work recovered during the viscoelastic relaxation represents the work that is needed to overcome the hysteresis friction, which is also termed as the deformational friction.

Consider a long cylinder sliding smoothly over a viscoelastic substrate at the velocity V under an applied force per unit length P. This movement creates a contact zone $-x_1 < x < x_2$, which divides into two parts with x_2 in the front and x_1 in the back. The size $x_1 + x_2$ and the shape x_1/x_2 of the contact characterize the hysteresis.

Before examining the friction, let us review a few basics of viscoelasticity. Boltzmann superposition principle is one starting point for the integral representation of linear viscoelasticity presented in chapters 4 and 6. An equally valid starting point is to relate the normal surface pressure p and strain e by a linear differential equation as a differential representation. The most general form is

$$\sum_i a_i \frac{d^i p}{dt^i} = \sum_i b_i \frac{d^i e}{dt^i}.$$ (8.80)

Because $dx/dt = -V$, the above equation becomes

$$Lp(x) = b_0 Q(x),$$ (8.81)

with the linear differential operator L given by

$$L = \sum_i a_i(-V)^i \frac{d^i}{dx^i}$$ (8.82)

and

$$Q(x) = \frac{1}{b_0} \sum_i b_i(-V)^i \frac{d^i e(x)}{dx^i}.$$ (8.83)

The surface displacement in the vicinity of contact is

$$z(x) = z_0 - Z(x),$$ (8.84)

where $Z(x)$ describes the shape of the slider and z_0 is the depth of penetration at $x = -s$, with $s = (x_2 - x_1)/2$, which goes to zero for static contact. When the viscoelasticity of the substrate is taken into account, the modified shaped of the slider is

$$\varsigma(x) = \frac{1}{b_0} \sum_i b_i(-V)^i \frac{d^i Z(x)}{dx^i} \equiv \varsigma_0 + \varsigma_s + \varsigma_a,$$ (8.85)

where ς_0 is a constant and ς_s and ς_a are the symmetrical and asymmetrical terms.

Generalizing the integral solutions for the elastic contact, we obtain

$$Q_0(x) = -\frac{\varsigma_0}{\ln 2 \cdot (l^2 - x^2)^{1/2}},$$ (8.86)

$$Q_s(x) = \frac{1}{\pi}(l^2 - x^2)^{1/2} \int_0^l \frac{y\varsigma_s'(y)\,dy}{(y^2 - x^2)(l^2 - y^2)^{1/2}},$$ (8.87)

and

$$Q_a(x) = \frac{2}{\pi}\frac{d}{dx} \int_x^l \frac{y\,dy}{(y^2 - x^2)^{1/2}} \int_0^y \frac{\varsigma_a'(\rho)\,d\rho}{(y^2 - \rho^2)^{1/2}},$$ (8.88)

where $l = (x_1 + x_2)/2$ is the semicontact length. Eqs. (8.86) and (8.87) follow from [24] and Eq. (8.88) from [25]. The summation of Q_0, Q_s, and Q_a gives $Q(x)$. When $Q(x)$ is substituted into Eq. (8.81), the formal solution for $p(x)$ can be obtained easily. The contact problem has to satisfy

$$p(-l) = 0$$ (8.89)

and

$$\int_{-l}^{l} p(x)\,dx = P.$$ (8.90)

Consider a simple viscoelastic solid that is frequently used in the literature for illustrative purpose,

$$p - V\tau_1\frac{dp}{dx} = E_\infty\left(e - V\tau_2\frac{de}{dx}\right),$$ (8.91)

where E_∞ is the relaxed modulus [see Eq. (6.17)], and τ_1 and τ_2 are the relaxation and retardation times, respectively. The unrelaxed modulus E_0 is related to the three independent parameters by $E_0/E_\infty = \tau_2/\tau_1 \equiv \varphi > 1$. The shape of the slider is $Z(x) = (x + s)^2/2R$, where R is the radius. Combining this equation with Eq. (8.85) results in

$$\varsigma(x) = Z(x) - V\tau_2 Z'(x) = \frac{s(s - 2V\tau_2)}{2R} + \frac{s - V\tau_2}{R}x + \frac{x^2}{2R}.$$ (8.92)

Substituting Eq. (8.92) into eqs. (8.86)–(8.88) yields

$$Q(x) = \frac{l}{R}\left\{\sqrt{1 - (x/l)^2} - \frac{1}{\sqrt{1 - (x/l)^2}}\left[\frac{s(s - 2V\tau_2)}{2l^2 \ln 2} + \frac{s - V\tau_2}{l} \cdot \frac{x}{l}\right]\right\}.$$ (8.93)

Solving Eq. (8.91), we obtain the contact pressure

$$p(x) = \frac{E_\infty}{V\tau_1}\int_x^l Q(y)\exp\left(\frac{x - y}{V\tau_1}\right)dy.$$ (8.94)

The effects of the sliding speed and the hysteresis on the deformational friction can now be evaluated from these two equations.

Substituting eqs. (8.93) and (8.94) into Eq. (8.89) gives

$$s/l_\infty = c + V\tau_2/l_\infty - \sqrt{(c + V\tau_2/l_\infty)^2 - 2c(V\tau_2/l_\infty)(1 - 1/\varphi)}, \qquad (8.95)$$

where $l_\infty = \sqrt{2RP/\pi E_\infty}$ is the Hertz semicontact length at the relaxed state and $c = \ln 2 \cdot (l/l_\infty)[I_0(l\varphi/V\tau_2)/I_1(l\varphi/V\tau_2)]$. Here, I_0 and I_1 are the modified Bessel functions of the zeroth and first orders, respectively. Similarly, Eq. (8.90) leads to

$$l/l_\infty = \sqrt{1 - \frac{s}{l_\infty \cdot \ln 2}\left(2\frac{V\tau_2}{l_\infty} - \frac{s}{l_\infty}\right)}. \qquad (8.96)$$

Solving eqs. (8.95) and (8.96) simultaneously gives s/l_∞ and l/l_∞ in terms of the nondimensional parameters $V\tau_2/l_\infty$ and φ.

The frictional coefficient is determined as the ratio of the horizontal force to the normal load from eqs. (8.93)–(8.96):

$$f = \frac{1}{RP}\int_{-x_1}^{x_2} xp(x)\,dx = \frac{l_\infty}{R}\left\{\frac{s}{l_\infty}\left[1 - \left(\frac{l}{l_\infty}\right)^2\right] + \frac{V\tau_2}{l_\infty}\left[\left(\frac{l}{l_\infty}\right)^2 - \frac{1}{\varphi}\right]\right\}. \qquad (8.97)$$

As the nondimensional speed of the slider increases, the reduction in the size of contact, the asymmetry in the shape of contact, and the corresponding variation of the friction coefficient are calculated in Figure 8.11. The symmetry contact

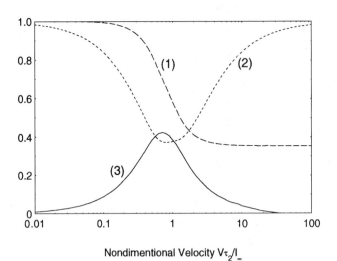

Nondimentional Velocity $V\tau_2/l_\infty$

FIGURE 8.11. Dependence of the hysteresis and friction on the sliding speed and material parameters. Curve 1 is the contact length (l/l_∞), curve 2 is the contact asymmetry (x_1/x_2), and curve 3 is the deformational friction (fR/l_∞). The non-dimensional parameters are plotted with the viscoelastic parameter $\varphi = 8$.

$x_1/x_2 = 1$ is reached when $V\tau_2/l_\infty \to 0$ or ∞. Interestingly, this figure reveals that all of the sudden changes in the size and shape of the contact and the deformational friction occur near $V\tau_2/l_\infty = 1$. In a dynamic viscoelastic experiment, the complex modulus is $E(\omega) = E'(\omega) + iE''(\omega)$, where ω is the angular frequency. If we assume $V/l_\infty = \omega$ in the contact problem, the loss tangent of Eq. (8.91) reaches its maximum

$$\tan \Delta_{max} = \left.\frac{E''}{E'}\right|_{max} = \frac{E_0 - E_\infty}{2\sqrt{E_0 E_\infty}} \tag{8.98}$$

at $V\tau_2/l_\infty = \omega\tau_2 = 1$. The peak of the loss tangent is a clear indication of the glass transition, where the greatest change in l/l_∞, the minimum value of x_1/x_2, and the maximum value of friction take place.

We are now going to explore whether the general theory of surfaces in sliding contact presented here could be used in a theoretical interpretation of the stick–slip transition observed recently in the sliding of atomically smooth solids. In this sort of slip regime, the forces are predominantly dissipative and could be identified with dynamic friction. The energy dissipated per unit time and unit area from the sliding contact is $E_{dis} = f\overline{p}V$, where $\overline{p} = P/2l$ is the mean pressure. By using eqs. (8.96) and (9.97), the nondimensional energy dissipated is calculated in Figure 8.12, which looks like the measured stick–slip transition (see Figure 3 in [26], where the deformation amplitude is proportional to the sliding velocity). This transition becomes steeper as the ratio φ of the unrelaxed modulus to the relaxed modulus gets larger. Although the dissipative energy continues to increase

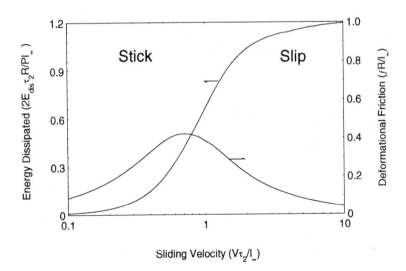

FIGURE 8.12. Transition with increasing sliding speed is illustrated in terms of the energy dissipated per unit time and the deformational friction. All coordinates are non-dimensional. $\varphi = 8$ is assumed in the calculation.

at a slower rate after passing this dynamic transition, the coefficient of friction increases in the stick regime but drops in the slip regime. Our calculation reveals that this kind of dynamic phase transition of soft matter in a tight spot depends on the contact geometry and viscoelastic materials that may be modified by the glass transition (see chapters 5 and 6). Consider $\tau_2 \sim 10^{-2}$ sec and $\ell_\infty \sim 1$ nm, the critical sliding velocity is $V_c \sim 10^{-5}$ cm/sec.

8.9 Diffuse Scattering

The physics of light scattering from rough surfaces should serve as the fundamental basis of understanding not only the surface structures, but also interesting problems, like the gloss and color [27] of images formed on printed documents. High gloss results from the specular reflection of light from a smooth surface. When the surface becomes rougher, the gloss is reduced and the diffuse scattering increases shown in Figure 8.13 [28]. The specular wave is often termed the coherent field because of its predictable phase relative to the incident wave. The diffuse wave is termed incoherent field, because its wide angular spread and lack of phase relationship with the incident wave.

Consider a monoenergetic incident wave vector $\vec{q}_o$ is scattered quasi-elastically at point $\vec{r}_o$ on a rough surface. The amplitude of the scattered spherical wave at

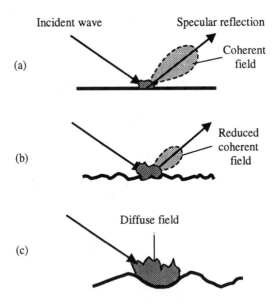

FIGURE 8.13. The transition of the light scattering from surfaces of different roughness. The increasing diffuse field and decreasing coherent field are depicted as roughness increases.

point $\vec{r}_s$ having a wave vector $\vec{q}_s$ is

$$a \sim \frac{1}{r_s} \exp[i(\vec{q}_o \cdot \vec{r}_o + \vec{q}_s \cdot \vec{r}_s)]$$

$$\cong \left[\frac{\exp(i\vec{q}_s \ell)}{\ell} \right] \exp(-i\vec{q} \cdot \vec{r}_o) \sim \exp(-i\vec{q} \cdot \vec{r}_o), \qquad (8.99)$$

where $\vec{q} = \vec{q}_s - \vec{q}_o$ is the momentum transfer vector and ℓ is the distance of a detector of scattered wave from the origin. ℓ is assumed much larger than $|\vec{r}_o|$. For a continuous surface, Eq. (8.99) can be generalized to have the form

$$a(\vec{q}) = \int \rho(\vec{r}, z') \exp(-i\vec{q} \cdot \vec{r}) \, dz' d^2 r. \qquad (8.100)$$

Here, ρ is the height distribution function mentioned in Eq. (8.1). Consider $\vec{q} = \vec{q}_r + \vec{q}_z$, where $\vec{q}_r$ and $\vec{q}_z$ are the components of the wave vector parallel and perpendicular to a smooth reference surface, respectively. Using this process and Eq. (8.100), we obtain the intensity of scattered light

$$I(\vec{q}) \sim \langle a(\vec{q})a^*(\vec{q}) \rangle = \int d\vec{r} \langle \exp[iq_z \Delta h(\vec{r})] \rangle \exp(i\vec{q}_r \cdot \vec{r})$$

$$= \int d\vec{r} \exp\left[-\frac{q_z^2}{2} C(\vec{r}) \right] \exp(i\vec{q}_r \cdot \vec{r}). \qquad (8.101)$$

The intensity $I(\vec{q})$ is a measure of the surface reflectance and may also be interpreted as the structure factor. In Eq. (8.101), $C(\vec{r})$ is the height correlation function given by Eq. (8.14). Hence,

$$I(\vec{q}) \sim \int_0^{\infty} \exp\left[-\frac{q_z^2}{2} C(r) \right] r \, dr \int_0^{2\pi} \exp(iq_r r \cos\varphi) \, d\varphi$$

$$= 2\pi \int_0^{\infty} \exp\left\{ -\frac{q_z^2 \sigma^2}{2} [1 - e^{-(r/\xi)^{2\alpha}}] \right\} J_0(q_r r) r \, dr, \qquad (8.102)$$

where $J_0(q_r r)$ is the Bessel function of the first kind of order zero. This equation establishes the basic relationship between the scattered intensity and the roughness parameters σ, ξ, and α, which defines the surface microstructure. Both specular and diffuse scatterings are included in this general expression.

The specular reflection I_{sp} occurs when $q_r = 0$, which leads to the Debye–Waller type of surface reflectance [28]:

$$I_{sp}(q_z, q_r = 0) = \exp\left(-q_z^2 \sigma^2/2\right) \equiv \Theta, \quad \text{for } r \gg \xi. \qquad (8.103)$$

When $r \ll \xi$, Eq. (8.102) approximates as

$$I_{sp}(q_z, q_r = 0) \sim \frac{2\pi \xi^2}{(q_z \sigma)^{2/\alpha}} \int_0^\infty \exp\left(-\tfrac{1}{2}x^{2\alpha}\right) x \, dx$$

$$= \frac{\pi 2^{1/\alpha} \xi^2}{\alpha (q_z \sigma)^{2/\alpha}} \Gamma(1/\alpha), \qquad 0 < \alpha \leq 1, \qquad (8.104)$$

where Γ is the gamma function.

Using Eq. (8.103) and noting

$$\int d\vec{r} \exp(i\vec{q}_r \cdot \vec{r}) = (2\pi)^2 \delta(\vec{q}_r), \qquad (8.105)$$

we separate the total light intensity given by Eq. (8.101) into two components: a sharp central delta-contribution and a broad diffuse contribution [29]:

$$I(\vec{q}_r, \vec{q}_z) = \Theta \cdot \delta(\vec{q}_r) + I_{diff}(q_r, q_z). \qquad (8.106)$$

The diffuse scattering is

$$I_{diff}(q_r, q_z) \sim 2\pi \xi^2 \Theta \int_0^R \left[\exp\left(\frac{q_z^2 \sigma^2}{2} e^{-x^{2\alpha}}\right) - 1 \right] J_o(q_r \xi x) x \, dx. \qquad (8.107)$$

When $r/\xi \to \infty$, $I_{diff}(q_r, q_z) \to 0$, which suggests that the diffusive scattering is short ranged. This method provides another consistent check that the short-range behavior characterizes both the microstructure and the diffuse light scattering of rough surfaces. By expanding the Bessel function

$$J_o(y) = \sum_{n=0}^\infty \frac{(-1)^n y^{2n}}{2^{2n} \Gamma^2(n+1)},$$

the asymptotic solution of Eq. (8.102) for $r \ll \xi$ is

$$I(q_r, q_z) \sim 2\pi \xi^2 \int_0^\infty \exp\left(-\frac{q_z^2 \sigma^2}{2} x^{2\alpha}\right) J_o(q_r \xi x) x \, dx$$

$$= \frac{\pi 2^{1/\alpha} \xi^2}{\alpha (q_z \sigma)^{2/\alpha}} \sum_{n=0}^\infty \frac{(-1)^n 2^{n(1/\alpha - 2)} (q_r \xi)^{2n} \Gamma\left(\frac{n+1}{\alpha}\right)}{(q_z \sigma)^{2n/\alpha} \Gamma^2(n+1)}, \qquad 0 < \alpha \leq 1.$$

$$(8.108)$$

The influence of the microstructure of rough surfaces on the light scattering is now explicitly shown in this equation. Therefore, Eq. (8.108) may serve as a useful tool for measuring the roughness parameters (σ, ξ, α) by light scattering techniques.

8.10   Surface Growth

In the deposition or growth phenomena, the surface structure is time dependent, and this time variation has been simulated numerically [1]. Consider a random growth process starting from a flat surface at $t = 0$. The two competing physical mechanisms in the formation of a growing surface are the vertical roughening and the lateral smoothing, which are driven by the random fluctuations coupled with surface diffusion. Let us begin by investigating the influence of both time and space on the surface structure in accordance with the dynamic scaling. In addition to Eq. (8.6), the time-invariant transformation,

$$t \rightarrow m^\kappa t, \tag{8.109}$$

is needed for a complete description of a time-dependent self-affine rough surface, where κ is the dynamic exponent. The interfacial width, which plays the role of a correlation length in the growth direction, should be scaled as a generalized homogeneous function,

$$\sigma(mL, m^\kappa t) = m^\kappa \sigma(L, t), \tag{8.110}$$

where L is the lateral size of the sample surface. When $m = 1/L$, the above equation becomes

$$\sigma(L, t) = L^\alpha F\left(\frac{t}{L^\kappa}\right), \tag{8.111}$$

where F is the scaling function. Two different scaling regions are dependent on the argument t/L^κ, as follows.

(1) At small time or long length scale, the scaling function increases as a power law:

$$F(t/L^\kappa) \sim (t/L^\kappa)^{\alpha/\kappa}, \quad \text{for } t/L^\kappa \ll 1. \tag{8.112}$$

Eqs. (8.111) and (8.112) give

$$\sigma(L, t) \sim t^{\alpha/\kappa}, \tag{8.113}$$

where α/κ can be called the growth exponent. The correlation length parallel to the surface is

$$\xi(t) \sim t^{1/\kappa}. \tag{8.114}$$

The corresponding time-dependent height correlation function is

$$C(L, t) \sim t^{2\alpha/\kappa}, \quad \text{for } \xi \ll L. \tag{8.115}$$

(2) As $t \rightarrow \infty$, the width saturates and the growth evolves into a steady state for which no characteristic length scale is below L. In this limit, we have

$$F(t/L^\kappa) \sim const, \quad \text{for } t/L^\kappa \gg 1 \tag{8.116}$$

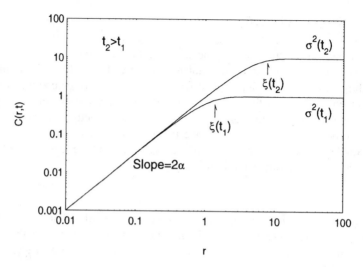

FIGURE 8.14. Schematics of the time-dependent height correlation function of a growing surface described on the basis of the dynamic scaling.

and

$$\xi \sim L. \tag{8.117}$$

Eqs. (8.111) and (8.116) give

$$\sigma_{sat}(L) \sim L^{\alpha}. \tag{8.118}$$

This equation shows that the saturation width increases with the system size. Thus, the saturation phenomena constitute a finite size effect. Figure 8.14 summarizes what has been discussed about the main features of the dynamic scaling.

The surface growth models to be discussed in the following text will not only confirm the dynamic scaling laws, but also shed light on the values of the exponents. The first model to study the growth of interfaces by particle deposition was introduced by Edwards and Wilkinson [30]. The fluctuations of an interface is governed by

$$\frac{\partial h(\vec{r}, t)}{\partial t} = D_s \nabla^2 h + \mu(\vec{r}, t). \tag{8.119}$$

Here, D_s is associated with a surface diffusion (see Section 3.6), and the $D_s \nabla^2 h$ term tends to smooth the interface. The noise term $\mu(\vec{r}, t)$ incorporates the stochastic character of the fluctuation process. When the noise has no correlation between time and space, one has

$$\langle \mu(\vec{r}, t)\mu(\vec{r}', t') \rangle = \chi \delta^d(\vec{r} - \vec{r}')\delta(t - t'). \tag{8.120}$$

Substituting eqs. (8.6) and (8.109) into Eq. (8.119), and using a general property of the delta function

$$\delta^d(m\vec{r}) = \frac{1}{m^d}\delta(\vec{r}),$$
(8.121)

we find

$$m^{\alpha-\kappa}\frac{\partial h}{\partial t} = D_s m^{\alpha-2}\nabla^2 h + m^{-(d+\kappa)/2}\mu.$$

Multiplying both sides of the above equation by $m^{\kappa-\alpha}$ yields

$$\frac{\partial h}{\partial t} = D_s m^{\kappa-2}\nabla^2 h + m^{(\kappa-d)/2-\alpha}\mu.$$
(8.122)

Because the Edwards and Wilkinson equation has to be invariant under the transformation from Eq. (8.119) to Eq. (8.122), each term on the right-hand side of Eq. (8.122) has to be independent of m, which results in $\kappa = d + 2\alpha = 2$. Therefore, the roughness, growth, and dynamic exponents are

$$\alpha = \frac{2-d}{2}, \quad \frac{\alpha}{\kappa} = \frac{2-d}{4}, \quad \text{and} \quad \kappa = 2,$$
(8.123)

respectively. From this process, we get $\alpha = 1/2$ for $d = 1$, $\alpha = 0$ for $d = 2$, and α becomes negative for $d > 2$, which means the interface is flat. Every noise-induced irregularity is softened by the surface diffusion.

Alternatively, the scaling function and its exponents can be solved exactly from the linear growth equation (8.119) via the Fourier transform in space and time,

$$h(\vec{q}, \omega) = \frac{\mu(\vec{q}, \omega)}{D_s q^2 - i\omega},$$
(8.124)

where $h(\vec{q}, \omega)$ is the Fourier transform of $h(\vec{r}, t)$. This method leads to the correlation function

$$\langle h(\vec{q}, \omega)h(\vec{q}', \omega')\rangle = \frac{\langle \mu(\vec{q}, \omega)\mu(\vec{q}', \omega')\rangle}{(D_s q^2 - i\omega)(D_s q'^2 - i\omega')}.$$
(8.125)

By using Eq. (8.120), the Fourier inversion of the above equation results in [1]

$$\langle h(\vec{r}, t)h(\vec{r}', t')\rangle = \frac{\chi}{D_s}|\vec{r} - \vec{r}'|^{2-d}F\left(\frac{D_s|t - t'|}{|\vec{r} - \vec{r}'|^2}\right).$$
(8.126)

Here, $F(u) \rightarrow e^{(2-d)/2}$ as $u \rightarrow 0$, and $F(u) \rightarrow const$ as $u \rightarrow \infty$. Comparing eqs. (8.111) and (8.126), we obtain the same scaling exponents shown in Eq. (8.123).

To include lateral growth into the growth equation, we add a new particle to the surface. The growth occurs locally, normal to the interface at the velocity v, generating an increase δh along the h axis that is

$$\delta h = \sqrt{(v\,\delta t)^2 + (v\,\delta t \cdot \nabla h)^2} = v\,\delta t[1 + (\nabla h)^2]^{1/2} \approx v\,\delta t\left[1 + \tfrac{1}{2}(\nabla h)^2 + \cdots\right]$$

for $|\nabla h| \ll 1$. This process suggests that a nonlinear term of this form has to be presented in the growth equation. Adding this term to Eq. (8.119), we obtain the Kardar, Parisi, and Zhang equation [31]:

$$\frac{\partial h(\vec{r}, t)}{\partial t} = D_s \nabla^2 h(\vec{r}, t) + \frac{v}{2}[\nabla h(\vec{r}, t)]^2 + \mu(\vec{r}, t). \tag{8.127}$$

This equation has been extensively studied in the literature [1]. It has been shown that in the strong coupling limit of large v the exponents α and κ are related by the simple equation $\alpha + \kappa = 2$.

Appendix 8A   Surface Forces

We would like to look closer at the attractive forces in their relationships with the wetting and adhesion of smooth surfaces under equilibrium conditions. One of the basic molecular forces acting at the interface is the dispersion force that involves the simultaneous excitation of molecules. We shall restrict ourselves to nonpolar media and deal only with van der Waals forces. The interaction energy of two molecules caused by dispersion force can in be written as

$$E_{dm}(r) = -\Lambda_{12}/r^6, \tag{8A-1}$$

where Λ_{12} is a constant and r is the distance between the two molecules. Granted additivity, and that the bodies contain molecules at the number densities N_1, N_2, the total energy per unit volume of interaction will be given by

$$E_d = -\int_{V_1} dv_1 \int_{V_2} dv_2 \frac{\Lambda_{12} N_1 N_2}{r_{12}^6}, \tag{8A-2}$$

where r_{12} is the distance between the volume elements dv_1 and dv_2 of bodies 1 and 2 with volumes V_1 and V_2. By allowing for repulsive forces shown in Figure 8A-1, an atomic-level separation z_c (much less than a manometer) between the two macroscopic bodies is reached at equilibrium. Introducing the Hamaker constant $B_{12} = -\pi^2 N_1 N_2 \Lambda_{12}$, the pressure between two semi-infinite plates is [32]

$$p_d = -\nabla E_d(z) = -\frac{B_{12}}{6\pi z^3}. \tag{8A-3}$$

If we want to separate these two bodies from the critical separation z_c to infinity, we can calculate the work done to create two new surfaces in place of the interface. Hence,

$$\int_{z_c}^{\infty} p_d\,dz = \frac{B_{12}}{12\pi z_c^2} = w_{ad} \equiv \gamma_1 + \gamma_2 - \gamma_{12}, \tag{8A-4}$$

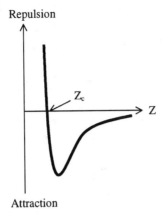

FIGURE 8A-1 Sketch of the interaction energy as a function of the surface separation between two flat surfaces via an attractive van der Waals force and a repulsive force where z_c is the critical separation.

where w_{ad} is the work of adhesion, γ is the surface energy, and the subscripts refer to bodies and interface. The Young equation has also been incorporated in Eq. (8A-4).

An alternative approach to calculating the interaction between any two bodies is based on the electomagnetic fluctuations developed by Lifshitz [33]. The van der Waals pressure between two semi-infinite plates is found to be

$$p_d = \frac{\hbar\overline{\omega}}{8\pi^2 z^3},$$ (8A-5)

where $\hbar$ is the Planck constant and the average angular frequency is

$$\overline{\omega} = \int_0^\infty \left[\frac{\varepsilon_1(i\zeta) - 1}{\varepsilon_1(i\zeta) + 1}\right]\left[\frac{\varepsilon_2(i\zeta) - 1}{\varepsilon_2(i\zeta) + 1}\right] d\zeta.$$ (8A-6)

According to the Kramers–Kronig relations, $\varepsilon(i\zeta)$ is related to the imaginary part $\varepsilon''(\omega)$ of the complex dielectric constant:

$$\varepsilon(i\zeta) = 1 + \frac{2}{\pi}\int_0^\infty \left[\frac{\omega\varepsilon''(\omega)}{\omega^2 + \zeta^2}\right] d\omega.$$ (8A-7)

Eqs. (8A-5)–(8A-7) reveal that the attraction is particular strong between bodies with large $\varepsilon''(\omega)$, which means strong absorption at large angular frequencies (far-UV region).

By following eqs. (8A-3)–(8A-5), the relations between the Hamaker constant, the Lifshitz–van der Waals constant $\hbar\overline{\omega}$, and the Dupre equation expressed by the equilibrium contact angle [see Eq. (8.34)] are established as

$$w_{ad} = \gamma(1 + \cos\theta_o) = \frac{B_{12}}{12\pi z_c^2} = \frac{\hbar\overline{\omega}}{16\pi^2 z_c^2}.$$ (8A-8)

References

1. A.-L. Barabasi and H. E. Stanley, *Fractal Concepts in Surface Growth* (Cambridge Univ., New York, 1995).
2. M. Marsili, A. Maritan, F. Toigo, and J. R. Banavar, Rev. Mod. Phys. **68**, 963 (1996).
3. T. Halpin-Healy and Y.-C. Zhang, Phys. Rep. **254**, 215 (1995).
4. J.-F. Gouyet, M. Rosso, and B. Sapoval, in *Fractals and Disordered Systems*, A. Bunde and S. Havlin, Eds. (Springer-Verlag, Berlin, 1991).
5. R. Chiarello, V. Panella, J. Krim, and C. Thompson, Phys. Rev. Lett. **67**, 3408 (1991).
6. T. S. Chow, Phys. Rev. Lett. **79**, 1086 (1997).
7. J. Feder, *Fractals* (Plenum, New York, 1989).
8. L. P. Kadanoff, Phys. Today, **36** (12), 46 (1983).
9. L. D. Landau and E. M. Lifshitz, *Fluid Mechanics* (Addison-Wesley, Reading, MA, 1959).
10. J. F. Joanny and P. G. de Gennes, J. Chem. Phys. **81**, 552 (1984).
11. P. G. de Gennes, Rev. Mod. Phys. **57**, 827 (1985); P. G. de Gennes, *Soft Interfaces* (Cambridge Univ., New York, 1997).
12. W. A. Zisman, in *Contact Angle, Wettability and Adhesion*, F. M. Fowkes, Ed. (Advances in Chemistry Series, No. 43, American Chemical Society, Washington, D.C., 1964).
13. R. N. Wenzel, J. Phys. Colloid Chem. **53**, 1466 (1949).
14. A. O. Parry, P. S. Swain, and J. A. Fox, J. Phys: Condens. Matter **8**, L659 (1996).
15. T. S. Chow, J. Phys: Condens. Matter **10**, L445 (1998).
16. B. W. Cherry and C. M. Homes, J. Colloid Interface Sci. **29**, 174 (1969).
17. H. Schoniorn, H. I. Frisch, and T. K. Kwei, J. Appl. Phys. **37**, 4967 (1966).
18. F. Ree, Teresa Ree, T. Ree, and H. Eyring, Advan. Chem. Phys. **4**, 1 (1962).
19. J. F. Douglas, H. E. Johnson, and S. Granick, Science **262**, 2010 (1993).
20. J. N. Israelachvili, *Fundamentals of Friction: Macroscopic and Microscopic Processes*, I. L. Singer and H. M. Pollock, Eds. (Kluwer, The Netherlands, 1992), p. 351.
21. L. D. Landau and E. M. Lifshitz, *Theory of Elasticity* (Pergamon, Oxford, 1959).
22. M. W. Pascoe and D. Tabor, Proc. Roy. Soc. A **235**, 210 (1956).
23. K. Tanaka, J. Phys. Soc. Japan **16**, 2003 (1961).
24. N. L. Muskhelishvili, *Some Basic Problems of Mathematical Theory of Elasticity* (Noordhoff, Groningen, 1953).
25. T. S. Chow, Wear **51**, 355 (1978).
26. S. Granick, Phys. Today **52** (7), 26 (1999).
27. F. W. Billmeyer, Jr., and M. Saltzman, *Principles of Color Technology* (Wiley, New York, 1981).
28. J. A. Ogilvy, *Theory of Wave Scattering from Random Rough Surfaces* (Institute of Physics Publishing, Bristol, 1991).
29. H.-N. Yang, T.-M. Lu, and G.-C. Wang, Phys. Rev. Lett. **68**, 2612 (1992).
30. S. F. Edwards and D. R. Wilkinson, Proc. R. Soc. (London) A **381**, 17 (1982).
31. M. Kardar, G. Parisi, and Y.-C. Zhang, Phys. Rev. Lett. **56**, 889 (1986).
32. J. Mahanty and B. W. Ninham, *Dispersion Forces* (Academic, New York, 1976).
33. E. M. Lifshitz, Soviet Phys. JETP **2**, 73 (1956).

Index